PEACEFUL AND NON-PEACEFUL USES OF SPACE

PEACEFUL AND NON-PEACEFUL USES OF SPACE

Problems of Definition for the Prevention of an Arms Race

Edited by

Bhupendra Jasani

UNIDIR
United Nations Institute for Disarmament Research

TAYLOR & FRANCIS
New York • Philadelphia • Washington, DC • London

USA	Publishing Office:	Taylor & Francis New York, Inc. 79 Madison Ave., New York, NY 10016-7892
	Sales Office:	Taylor & Francis Inc. 1900 Frost Road, Bristol, PA 19007-1598
UK		Taylor & Francis Ltd. 4 John St., London WCIN 2ET

PEACEFUL AND NON-PEACEFUL USES OF SPACE: Problems of Definition for the Prevention of an Arms Race

1 2 3 4 5 6 7 8 9 0   B R B R   9 8 7 6 5 4 3 2 1

Cover design by Debra Eubanks Riffe.
A CIP catalog record for this book is available from the British Library.

Library of Congress Cataloging-in-Publication Data

Peaceful and non-peaceful uses of space : problems of definition for
the prevention of an arms race / edited by Bhupendra Jasani.
p. cm.
Includes bibliographical references and index.

1. Space law. 2. Space weapons (International law) I. Jasani, Bhupendra.
JX5810.P43 1991
341.4'7—dc20 91-6936
ISBN 0-8448-1709-0 CIP

UNIDIR

United Nations Institute for Disarmament Research

Institut des Nations Unies pour la Recherche sur le Désarmement

UNIDIR is an autonomous institution within the framework of the United Nations. It was established in 1980 by the General Assembly for the purpose of undertaking independent research on disarmament and related problems, particularly international security issues.

The work of the Institute aims at:

1. Providing the international community with more diversified and complete data on problems relating to international security, the armaments race, and disarmament in all fields, particularly in the nuclear field, so as to facilitate progress, through negotiations, toward greater security for all States and toward the economic and social development of all peoples;

2. Promoting informed participation by all States in disarmament efforts;

3. Assisting ongoing negotiations on disarmament and continuing efforts to ensure greater international security at a progressive lower level of armaments; particularly nuclear armaments, by means of objective and factual studies and analyses;

4. Carrying out more in-depth, forward looking, and long-term research on disarmament, so as to provide a general insight into the problems involved, and stimulating new initiatives for new negotiations.

The contents of UNIDIR publications are the responsibility of the authors and not of UNIDIR. Although UNIDIR takes no position on the view and conclusions expressed by the authors of its research reports, it does assume responsibility for determining whether they merit publication.

UNIDIR

Palais des Nations
CH-1211 Geneva 10
Tel. (022) 734 60 11

Table of Contents

Preface

Research on the prevention of an arms race in outer space has been an important component of UNIDIR's programme of research for many years. In 1987 UNIDIR published *Disarmament: Problems related to Outer Space* which sought to identify the current military uses of outer space, the possibilities of space arms development and their implications, the nature of the existing legal regime regarding outer space and the proposals made by states for preventing an arms race in outer space.

UNIDIR was encouraged by the reaction to this research report and the fact that it was frequently quoted by delegations in international fora as well as being well received in the academic community. Our second research project in the field of outer space is a logical corollary to the first. Problems of definition confront the negotiator at the very outset of any discussion on preventing an arms race in outer space. For example there is an obvious need for international consensus on the definition of a space weapon. Thus the need to pool the expertise of an international group of legal and scientific scholars to analyze these problems and offer optional approaches to the international community was considered by UNIDIR to be a useful research project.

We were fortunate to be able to assemble an eminent group of experts from different regions of the world comprising Khairy Aly, Raul Boix Amat, Yves Boyer, Ashton Carter, S. Chandrashekar, Chen Zhongqing, Pal Dunay, Hubertus Feigl, Dieter Felske, Donald Lee Hafner, the late Thomas Johnson, Stanislav Rodionov, Isabelle Sourbès, Paul Stares, Ivan Vlasic and Caesar Voûte. Bhupendra Jasani co-ordinated the project as a UNIDIR consultant and is the editor of the publication resulting from it. Some members of the group responded to our request to write Chapters on specific issues while others offered their comments and advisory opinions at various stages of the project. The members of the group, who functioned in their personal capacities and not as representatives of their governments or organizations, met three times during the course of the project. I thank them for their valuable contribution. I would also like to acknowledge the useful advice provided by Nandasiri Jasentuliyana, Chief, Outer Space Affairs Division, United Nations, New York and his staff during the course of the research project.

This publication is therefore the cumulative result of a long period of research by highly qualified experts in an attempt to describe the legal and scientific problems of arriving at definitions in the task of preventing an arms race in outer space. It is our hope that it will assist the international community in the active discussion of these issues that is going on.

Although UNIDIR customarily takes no position on the views and conclusions expressed by individual authors of its research publications it does assume responsibility for determining whether research reports merit publication. Consequently we commend this report to the attention of its readers.

Jayantha Dhanapala
Director, UNIDIR

Chapter 1: Introduction

Bhupendra Jasani

During the 1980s considerable attention was focused on activities in outer space. This involved both the civil or the "peaceful" and military uses of space. However, no progress in the field of arms control in outer space was made, particularly on a multilateral basis. (The creation of the *Ad Hoc* Committee on the Prevention of an Arms Race in Outer Space, a subsidiary body of the Conference on Disarmament in Geneva, was only a procedural step.) This is probably because, at best, many of the relevant issues are not clearly defined in the existing space law.

For example, the first treaty that limited activities in outer space was the 1963 Treaty banning nuclear weapon tests in the atmosphere, in outer space and under water. Here the terms "atmosphere" and "outer space" are used without defining where the atmosphere ends and outer space begins. Again the 1967 Treaty on principles governing the activities of States in the exploration and use of outer space, including the Moon and other celestial bodies is silent on the meaning of "outer space". Thus, the first task of this research report was, at least, to indicate how this demarcation could be drawn.

The 1967 Outer Space treaty in its preamble speaks about the "use of outer space for peaceful purposes." Moreover, in article IV paragraph two, it is stated that "The Moon and other celestial bodies shall be used ... exclusively for peaceful purposes." Yet the term "peaceful purposes" is not defined in the treaty. It is now well recognized that military uses of outer space consist mainly of a large number of satellites to enhance the capabilities of terrestrial weapons as well as those of space weapons. The use of remote sensing, civil communications and navigation, and scientific satellites are generally regarded as peaceful uses of outer space. However, many of these spacecraft, with higher capabilities, become military satellites. Not only this, but they became components of terrestrial weapon systems. Examples of these are the reconnaissance spacecraft for target identification, military communications satellites for command and control functions and navigation satellites to guide weapons accurately to their targets. Thus, the notion of equating "peaceful" with "non-aggressive" was developed during the late 1960s. However, with increased capabilities of civil satellites, it is becoming increasingly difficult to draw a distinction between civil and military spacecraft. In the report, an attempt is made to look for technical parameters that could be used to distinguish between a peaceful and a military satellite. Another criterion might be to consider the functions and users. For example, if a satellite is used for verification of arms control agreements by an international verification agency, then such a spacecraft is less likely to be used for any aggressive purposes. Such concepts are explored in the present report.

From the above it can be seen that in order to define space weapons, it is useful to define not only its components such as satellites but also outer space itself in order to know when a weapon is a space weapon. As regards to satellites in this context, it is desirable to distinguish between peaceful and military spacecraft. Equipped with these parameters, the research report has attempted to suggest a definition for a space weapon and its components. A related question of non-dedicated systems that might be used as space weapons is also considered.

During the more than three decades of the space age (beginning in 1957), well over 3,000 satellites have been launched. More than 75 percent of these satellites have been on military missions. Moreover, during this period a number of weapons tests to destroy such satellites have been carried out. Tests of nuclear weapons have taken place and missiles to deliver such weapons through outer space to their targets have been developed, tested, and deployed on the surface of the earth. Such long-range missiles, even today, travel through outer space during their test flights. Of course, they do not carry their nuclear warheads. Yet outer space has remained free of weapons, that is, no weapon has been deployed in space.

The United States and the USSR have developed and tested and even deployed anti-ballistic missile (ABM) defense systems. However, soon after the treaty between the United States and the Union of Soviet Socialist Republics on the Limitation of Anti-ballistic Missile Systems (the 1972 ABM Treaty), the United States dismantled its one allowed system protecting its intercontinental ballistic missile (ICBM) silos. The Soviet Union, on the other hand, retained its system around Moscow.

Thus, with the availability of long-range missiles, which could reach altitudes of some 1,000 km during their ballistic trajectories, and ABM systems, the two superpowers have limited anti-satellite (ASAT) weapons capability also. Moreover, both are conducting research on anti-tactical ballistic missile (ATBM) defense systems, and they continue to explore the feasibility of advanced ASAT and ABM weapons. On the drawing boards in the laboratories of both the space powers exist futuristic arsenal such as sophisticated missiles and beam weapons. This situation has existed for most of the three decades plus during which the two superpowers dominated outer space.

In spite of this, the two space powers have indicated an interest in exploring arms control to regulate some military activities in outer space. This is seen by the current bilateral negotiations in the Geneva Defense and Space Talks (DST). However, these talks are dominated by bilateral issues. A successful outcome would legally protect space objects of the superpowers only.

Many more countries are beginning to use outer space for such non-military purposes as remote sensing, communications, and meteorology. The difference between the capabilities of remote sensing and military reconnaissance satellites is becoming so small that, to some extent, even the former could be used for military surveillance purpose. This would have an implication for the definition of "peaceful" uses of outer space. Observation satellites could be used for peacekeeping, crisis monitoring, and verification of arms control treaties, and yet they are used for targeting purposes. Whereas the former three are "peaceful" activities, they cause some uneasiness because satellites in any of the three roles observe military activities of states. One way of alleviating such anxieties is to put some qualitative limits on the capabilities of a satellite. The other is to internationalize verification processes. Both of these are discussed below and in Chapters 3, 5, and 6.

As the capabilities of civil and military satellites converge, it is possible that civil spacecraft could become targets for ASAT weapons. Even accidental damage in time of crisis could aggravate an already tense situation. Satellites are not protected by general international law or by any specific multilateral treaty. Unlike military reconnaissance spacecraft, damage to such civil remote sensing satellites may not be regarded as an attack on a country's national security assets. However, during crisis, such satellites could become very important.

A number of nations now have ICBM capabilities and some are even developing ATBMs and air defense systems. These at least have some common elements to space weapons and, therefore, could be converted into, for example, ASAT weapons. Thus the regulation of the use of outer space could not entirely be the concern of the U.S.-Soviet bilateral negotiations. Therefore, although multilateral arms control matters cannot be considered independently of the developments in the bilateral forum, as far as ASAT weapons are concerned, discussions on arms control would be incomplete without the participation of other nations. It is, therefore, welcomed that in the United Nations General Assembly and in the Conference on Disarmament (CD) in Geneva, multilateral discussions on space have started.[1] However, even after nearly a decade of talks in these fora and in spite of several nations having ambitious plans to use outer space for peaceful and defence purposes, the considerations have still concentrated on exploring already existing arms control measures rather than actually negotiating for a new treaty.

There is no doubt that for an international community effectively to contribute to regulating military activities in outer space, an understanding of the nature of the problem is essential. Among other things, a basic element of the problem is the characteristics of the military hardware and activities in outer space to be controlled. Thus it would greatly assist negotiations for a multilateral space-related treaty if a clear definition of a space weapon were available. When considering this, inevitably several questions arise. For example, where does air space end and outer space begin? What is meant by "peaceful" and "non-peaceful" uses of outer space? What is a space object or a component? A number of arms control agreements relevant to military activities in outer space already exist in which weapons, components, and objects in space are mentioned. Therefore, it is also useful to examine these.

This book, therefore, attempts, with the help of 20 experts (see List of Contributors and Consultants to the Book) to identify issues and approaches that negotiators could adopt when considering the control of space weapons. Considerable effort is devoted to suggest ways to define the following:

1. Boundary between air space and outer space.
2. Peaceful and non-peaceful uses of outer space.
3. Space weapons.
4. Components and subcomponents of space weapons.

---

[1] UN resolution 36/99, 9 December 1981; UN resolution 36/97C, 9 December 1981; and UN resolution 39/59, 12 December 1984.

5. Testing: laboratory testing, field testing, testing in an ABM mode or in an ASAT mode.
6. A "laboratory" testing, particularly when this might be in space.
7. Non-dedicated space weapons.
8. Internationalization as a means of distinguishing "peaceful" from "non-peaceful" uses of outer space; verification has been taken as an important example of this.

I - Problems of Definitions

In the formulation of the body of space law,[2] only a limited attempts is made to define some of the relevant terms. Among the multilateral treaties, only two define (crudely) space objects. For example, according to the 1972 Convention in International Liability for Damage Caused by Space Objects, "A launching State shall be absolutely liable to pay compensation for damage caused by its space object on the surface of the earth or to aircraft in flight" (article II). Moreover, "In the event of damage being caused elsewhere than on the surface of the earth to a space object of one launching States or to persons or property on board such a space object by a space object of another launching State, the latter shall be liable only if the damage is due to its fault or the fault of persons for whom it is responsible" (article III). A "space object" is not defined explicitly but it "... includes component parts of a space object as well as its launch vehicle and parts thereof" (article I.d). (The meaning of space objects is discussed in some detail in Chapter 5.)

Article VI of the 1976 Registration Convention states that "Where the application of the provisions of this Convention has not enabled a States Party to identify a space object... which may be of a hazardous or deleterious nature, other States Parties, including in particular States possessing space monitoring and tracking facilities, shall respond to the greatest extent feasible to a request by that State Party, ... in the identification of the object." Again, in this convention, the definition of a "space object" is treated in the same way as in the 1972 Liability Convention. It should be noted that a space object" ... may be of a hazardous or deleterious nature."

None of the agreements defines the meaning of outer space or space weapons. The treaty on Principles Governing the Activities of States in the Exploration and use of Outer Space, including the Moon and other Celestial Bodies (1967 Outer Space Treaty) merely states that the parties " ... undertake not to place in orbit around the earth any objects carrying nuclear weapons or any other kinds of weapons of mass destruction, ..." (article IV).

[2] The five multilateral treaties and conventions regulating activities of states in outer space are: (1) the 1967 Treaty on Principles Governing the Activities of States in the Exploration and Use of Outer Space, including the Moon and other Celestial Bodies, (2) the 1968 Agreement on the Rescue of Astronauts, the Return of Astronauts and the Return of Objects Launched into Outer Space, (3) the 1972 Convention on International Liability for Damage Caused by Space Objects, (4) the 1976 Convention on Registration of Objects Launched into Outer Space, and (5) the 1979 Agreement Governing the Activities of States on the Moon and other Celestial Bodies.

Although "weapons of mass destruction" have not been defined in the treaty, it is generally understood to include chemical and biological weapons also. The UN definition of weapons of mass destruction includes "radioactive material weapons." This means that a device, including any weapon or equipment, other than a nuclear explosive device, designed to cause destruction, damage, or injury by means of radiation from radioactive substances is a radiological weapon and a weapon of mass destruction. Thus the treaty prohibits a specific type of weapon in outer space without defining outer space.

In order to correct some of the above deficiencies, a number of States have proposed that essentially article IV of the Outer Space Treaty should be amended to include all weapons. For example, on 26 March 1979, Italy proposed an additional protocol to prohibit "the development and use of earth- or space-based systems designed to damage, destroy or interfere with the operations of other States' satellites."[3] In 1987 and again in 1988, Venezuela raised the question of amending article IV of the Outer Space Treaty.[4] The latter also suggested a definition of "space weapons."

There are considerable similarities between technologies used in strategic defense (known as ballistic missile defense -- BMD), air defense, anti-tactical ballistic missile (ATBM) defense and ASAT systems. Thus advances in many of these areas can boost development in the other areas. Therefore, the two bilateral treaties,[5] dealing with strategic defense and strategic arms limitation, are of interest for this book. For example, the 1972 ABM treaty defines ABM interceptor missiles, launchers, radars and ABM system components (article II). In the treaty, some quantitative parameters are also defined, for example, for radars (paragraph B of the Agreed Interpretations). The treaty also discusses testing of systems in an ABM mode (paragraph B of the Unilateral statements).

Because of the ASAT capability of ICBMs, it is useful to recognize that the 1972 Interim Agreement between the United States and the USSR on certain measures with respect to the limitation of strategic offensive arms (SALT I agreement) defines land-based ICBM launchers (paragraph A of the Agreed Interpretation). An ICBM launcher is defined in the SALT II agreement as "... launchers of ballistic missiles capable of ... a range in excess of 5,500 kilometres" (article II.1).

In the following discussion, an attempt is made to identify additional technical parameters that might be used to define space weapons. This includes the problems of defining outer space, "peaceful" and "non-peaceful" uses of this environment, and the question of non-dedicated space weapons.

---

[3] CD document CD/9, 26 March 1979.

[4] CD document CD/398, 19 March 1987 and CD/851, 2 August 1988.

[5] The two agreements are: (1) treaty between the U.S. and the USSR on the limitation of Antiballistic Missile systems (ABM treaty), and (2) Interim Agreement between the U.S. and the USSR on certain measures with respect to the limitation of Strategic Offensive Arms (SALT I agreement).

Where Does Outer Space Begin?

Linked with the definition of a space weapon is the definition of the boundary between air space and outer space. The 1967 Outer Space Treaty is silent on the meaning of the term "outer space." It is often argued that a definition is not needed because, for example, the treaty has worked without defining outer space (for details, see Chapter 2). This is true probably because the treaty does not contain significant arms control measures, and, until recently, space capabilities did not extend to reusable space vehicles and land- and space-based weapons for interceptions of satellites and missiles or their warheads travelling through space.

The United States space shuttle has made several flights in orbit (the first on 12 April 1981); the USSR tested its own reusable space plane called *Buran* on 15 November 1988 (an unmanned flight); the European Space Agency has serious plans to develop such vehicles; and Japan is likely to follow suit. An intense debate continues on what kind of legal regime is applicable to reusable space vehicles and the two superpowers actively continue with their space weapon programs, which make it important to devise a definition or at least some understanding of where outer space begins and air space ends.

For a number of years, the issue has been under consideration by the Legal Sub-Committee of the United Nations Committee on the Peaceful Uses of Outer Space (UNCOPUOS).[6] The debate has ranged from saying that a precise designation of the beginning of outer space is not necessary, or even inappropriate,[7] to saying that a more rigid definition is required. The latter has given rise to one definition that depends on some properties of the earth's atmosphere, which is divided into different layers (see Chapter 2). For example, the heterosphere, which consists of the troposphere, the stratosphere and the mesosphere, extends to 90, 100 km. At this altitude, the atmospheric drag is such that a spacecraft would be able to complete a full orbit around the earth without any further application of force. This is known as the spatial definition.

A second definition is based on a functional approach. In this case outer space would be defined according to various orbits used by satellites to perform a number of functions. For example, earth observations are made from low earth orbits, whereas communications between various points on the earth's surface are best carried out by satellites in the geostationary orbit, or GSO (for details, see Chapters 2 and 5).

Any particular approach adopted may depend on the type of arms control measure being discussed. If, for example, a future treaty on the prevention of arms in outer space requires an accurate demarcation between air and outer space for operational reasons, an arbitrary decision would have to be taken regarding the exact position of this boundary. Based on pragmatic consideration, an altitude of 100 km above the earth's surface could be

---

[6] The issue was first reported in the Report of the General Assembly ad hoc Committee on the Peaceful Uses of Outer Space, July 1959, and it has been formally put on the agenda of the Legal Sub-Committee of COPUOS since 1967.

[7] *Daily Bulletin*, U.S. Mission, Geneva/U.S. Embassy, Bern, No. 60, 11 April 1986.

a logical choice for such a demarcation. However, such a boundary need not have universal applicability for other cases where there exists a difference in the legal regime between air and outer space (see Chapter 2).

Some of the technical and legal aspects of the definition of boundary between the air and outer space are also discussed in some detail in Chapters 3 and 5.

"Peaceful" or "Non-Peaceful" Uses of Space

The 1967 Outer Space Treaty was the first international agreement dealing specifically with the conduct of nations in outer space. Among other measures is the Agreement Governing the Activities of States on the Moon and other Celestial Bodies (the 1979 Moon Treaty), which came into force in 1984. According it "The Moon shall be used by all States Parties exclusively for peaceful purposes" (article III.1). This was already a goal in the 1967 treaty (article IV). As for the near earth orbits, in the preamble of the 1967 Outer Space Treaty, only a vague recognition is made of "the common interest of all mankind in the progress of the exploration and use of outer space for peaceful purposes." Thus there seems to be a distinction between the ways in which outer space and the moon and other celestial bodies are to be used, and yet no treaty concerned with activities in outer space defines what is meant by "peaceful" and "exclusively for peaceful" purposes.

It was the United States that, in 1957, proposed that outer space should be used exclusively for scientific and peaceful purposes.[8] Along with Canada, France, and Britain, the United States also called for a report that looked into "the design of an inspection system which would make it possible to assure that the sending of objects through outer space will be exclusively for peaceful and scientific purposes".[9] The Soviet Union and its allies were opposed to both proposals in spite of considerable support from most of the UN members. It should be remembered that the USSR had already launched two artificial earth satellites, whereas the United States had none in orbit until February 1958. Since then, much discussion has taken place within the United Nations and elsewhere on the meanings of "peaceful" and "non-peaceful" uses of outer space, with many interpretations. However, in most multilateral agreements in which "peaceful" activities are considered, it is interpreted to mean "non-military" (as indicated in Chapter 3). Yet there are treaties, such as the Law of the Sea, in which the seabed is considered "non-military," whereas elsewhere in the treaty, by "peaceful" it is meant to be "non-aggressive."

A different interpretation is given by some emerging space powers to this issue. For example, it is suggested that "peaceful" should mean total absence of the military and force or conflict. It is argued that the present international legal regime on space does not

[8] President Dwight D. Eisenhower and UN Ambassador Henry Cabot Lodge, *Department of State Bulletin*, 36 (1957), 124, 227; and U.S. Secretary of States John Foster Dulles, *Department of State Bulletin*, 37 (1957), 271.

[9] *Department of State Bulletin*, 37 (1957), 453.

encourage this environment (see Chapter 5). Thus it is concluded that in future agreements the use of the word "peaceful" either be avoided altogether or defined clearly.

The early space activities of the two major space powers were largely devoted to orbiting satellites with various military missions. For example, a spacecraft orbiting close to the earth could identify potential military targets and determine their precise location. Satellites not only relayed military messages over long distances but could also guide modern missiles, aircraft, and naval vessels to their targets with considerable precision. The potential "peaceful uses" of some of these satellites cannot be disputed. For example, reconnaissance satellites have proved to be of enormous value in a number of different fields, including verification of arms control agreements. In this context, it might be suggested that a spacecraft used for verification of arms control agreements is a "peaceful" use of outer space if the sensors on board have resolution between 1 and 5 m. Resolution of sensors on board most of the intelligence-gathering satellites is better by a factor of about 10. Also, the altitude of a civil observation satellite tends to be about 800 km compared with 150 to 300 km for reconnaissance satellites. Another criterion is whether arms control verification data obtained from observation satellites is widely available (see section on verification later in this Chapter; also see Chapters 3 and 5).

Whether the use of navigation satellites is peaceful or not is also arguable. Past debates have not provided a clear definition, but the two superpowers appear to have agreed on "non-aggressive" to mean "peaceful." Under this definition, it is not clear whether military navigation satellites, which are used entirely to guide earth-based and in the future even space-based, weapons to unprecedented accuracies, would fall under this category. As in the case of the remote-sensing satellites, the performance requirements are different for civil and military uses.

Military satellites are components of weapon systems. Could they then be regarded as "non-aggressive"? In any case, the current definition of "peaceful" uses of space resulted in a proliferation of military satellites of the two major space powers. Now new states have emerged with such capabilities. Moreover, as the proliferation of missiles increases, so will the use of navigation and other military satellites, thereby creating a dilemma.

While arms control measures are being discussed, a number of confidence-building measures have been proposed. For example, measures such as "rules of the road," "keep out zones," "codes of conduct," and "immunity for satellites" have been suggested. As these are not strictly in the area of arms control, they may be referred to the UN Committee on the Peaceful Uses of Outer Space, or UNCOPUOS. It is, however, important to know how navigation satellites should be classified to determine whether they should come under the jurisdiction of UNCOPUOS. Until now, relatively little discussion has taken place on military satellites in this forum on the grounds that its mandate is to discuss "peaceful activities in outer space."

A more profound question is how tests of ASAT weapons in outer space should be regarded. These are clearly aggressive in nature and yet the superpowers have conducted a number of tests of their ASAT weapons and, no doubt, others may follow them. Such tests have created space debris, which could become harmful to the space activities of other nations. No mention of space debris is made in the 1967 Outer Space Treaty. However,

> If a States Party to the Treaty has reason to believe that an activity or experiment planned by it or its nationals in outer space, ... would cause potentially harmful interference with activities of other States Parties in the peaceful exploration and use of outer space, ... it shall undertake appropriate international consultations before proceeding with any such activity or experiment. A State Party to the Treaty which has reason to believe that an activity or experiment planned by another State Party in outer space, ... would cause potentially harmful interference with activities in the peaceful exploration and use of outer space, ... may request consultation concerning the activity or experiment (article IX, 1967 Outer Space Treaty).

Potentially harmful debris has been generated in outer space by the ASAT tests of the Soviet Union and the United States. In spite of this, as far as it is known, no "international consultations before proceedings with such activity" have been undertaken by either superpower. Nor have other states parties to the treaty found reasons to believe that such activities planned by the two superpowers might cause damage to their spacecraft from the debris generated by ASAT experiments. This is in spite of the fact that there is now some evidence that malfunction of a number of satellites may have occurred because they were hit by debris.

In time, as more and more nations begin to use space, the chances of harmful interference with their peaceful activities will increase. Again, without a clear definition of "peaceful" uses of space, who should discuss such issues? Should it be UNCOPUOS whose mandate does not include discussions on military activities in space, or the Conference on Disarmament (CD), which would not discuss debris because it is not an arms control issue? As more and more countries begin to use military satellites and even begin to test ASAT weapons, it will be necessary to come to some understanding on the definition of "peaceful" or "non-peaceful" uses of outer space.

The above discussion (and that in Chapters 3 and 5) focuses on the legal and political aspects of the notion of "peaceful," but Chapter 4 looks for technical parameters that might help to distinguish between the "peaceful" and "non-peaceful" uses of outer space. There are two approaches. One is to describe various satellites based on their technical orbital parameters, the nature of signals transmitted by then, their sizes and their behaviour in orbits. However, the capabilities of civil and military satellites are converging, making it increasingly difficult to distinguish between the two applications. Another approach is to examine military doctrines that dictate the use of force and then consider whether outer space contributes to the military requirements. However, such an approach may run into difficulties because most doctrines are based on the deterrence theory, military capabilities have to be effective to deter any aggression. The problem with this is an effective military force may be seen by an adversary as an offensive force. Thus the notion of "non-aggressive" use of outer space as allowable would run into difficulties. Moreover, as mentioned earlier, military satellites could be regarded as components of weapons on earth. These issues are discussed further in Chapters 3, 4, and 5.

Space Weapons

Broadly, space weapons can be divided in three groups: nuclear, non-nuclear weapons (see Table 1.1) and non-dedicated space weapons. The first two categories can be further classified according to their basing mode, into five types: ground-to-space, air-to-space, space-to-space, space-to-air, and space-to-ground (for different approaches to the definition of space weapons, see Chapter 5).

Nuclear explosive-based space weapons

In this category, only the ground-based BMD weapons are deployed for interception of missiles and their warheads above as well as within the atmosphere. Such weapons could also be used to destroy satellites in orbits (see Table 1.1). Conventional BMD or ASAT weapons, such as the Soviet system around Moscow, require rocket technologies to carry the warheads to the targets. The basic components of such systems are nuclear warheads, missiles, launchers/silos, and radars.

Advanced new generation nuclear explosive-based BMD weapons being investigated are the X-ray lasers. As the deployment of these in space is banned by the 1967 Outer Space Treaty, they would, if they worked, be ground- or air-based. Again, they would have a dual role as BMD and ASAT weapons.

Non-nuclear space weapons

These can be divided into two broad classes: kinetic energy (KEW) and directed energy (DEW) weapons.

Kinetic Energy Weapons. These are, for example, the traditional BMD systems using ICBM or SLABMs, which have the nuclear warheads replaced with computer-guided, non-nuclear explosives. The nuclear explosives are indiscriminate in their effects; they affect the intended targets as well as any other satellite or ICBM within the range of the nuclear blast. Moreover, nuclear effects interfere with other elements, such as detection and guidance radars, of the defense systems. In order to correct for this unpredictable aspect of the interceptors, extensive research is continuing on extremely accurate guidance systems. Advances in such devices are beginning to result in the replacement of nuclear devices with non-nuclear explosives.

Whereas the usefulness of ground-, air-, or space-based non-nuclear weapons as BMD systems is debatable, there is little doubt that such a system would be very effective as a long-range ASAT weapon. An ASAT system would consist of an interceptor that is launched into approximately the same orbit as the target satellite, maneuvers close to it, and then explodes, destroying the target with a shotgun-like blast of metal pellets. In another system the interceptor would destroy the target by direct collision with it (for details, see Chapter 6). The target could be either a satellite or a missile or its nuclear warhead. As an ASAT, the interceptor could also be launched from high-altitude aircraft.

Table 1.1. Space Weapons.

Weapons						Status
Type	Deployment mode[a]					
	Ground-space	Space-space	Space-air	Space-ground	Air-space	
NUCLEAR						
Endoatmospheric	X (1)					Existing
Exoatmospheric	0 (2)					Existing
X-ray laser	0	0 (3)	0 (3)	X (3,4)	0	Investigated
NON-NUCLEAR						
Projectiles (e.g., Kew, U.S./F-15 MHV, Soviet ground-based missiles, rail guns)[b]	0	0	0	0	0	Investigated; Soviet direct ascent reportedly deployed
Neutral particle beams		0				Investigated
Lasers	0 (5)	0	X (3)	X	0	Investigated
Radio-frequency weapons	0	0	0	0	0	Investigated

[a]Key:
X = ballistic missile defense only.
O = ballistic defense and ASAT capabilities.
1 = essentially ground to atmosphere.
2 = possible electromagnetic pulse weapon.
3 = 1967 Outer Space Treaty bans deployment.
4 = mainly for upper atmospheric applications.
5 = laser deployed preferably on top of mountains.
[b] MHV = miniature homing vehicle.

Another type of KEW, based on a new concept, is called the electromagnetic (EM) railgun. It is in some ways a hybrid between conventional weapons and more futuristic DEWs. This weapon involves the use of electromagnetic forces to accelerate a mass, much like an ordinary rifle or cannon uses the chemical reactions in gunpowder to accelerate a bullet or a cannon. The main difference between the EM railgun and conventional guns is that the EM forces can produce much higher speeds for projectiles.

Space mines and other uses of projectile satellites as ASATs are also being investigated. The fragile character of satellites makes them extremely vulnerable to collisions with objects, especially if an interceptor is moving against the target satellite (the relative speed between the two is much larger than the individual ones).

Directed Energy Weapons. In DEWs, instead of the destructive energy in the form of a mass moving at high speed, energy itself, travelling at the speed of light (300,000 km/s), is used to destroy the target. Essentially three types of DEWs are being investigated: particle beam weapons (PBW), high energy lasers (HEL), and radio-frequency weapons (RFW).

The PBWs have long been researched for possible military applications. Although charged particle beams are not very useful as weapons, at least as long-range weapons, because, among other things, they are deflected by the earth's magnetic field, neutral particle beams (NPBs) are being investigated for possible use in outer space as discriminators. The discrimination between nuclear warheads and decoys in the midcourse phase of their ballistic trajectory might be accomplished by irradiating them with the NPB and measuring the secondary radiation that is induced by the interactions between the NPB and the massive warheads. Thus such devices may become components of defensive space weapons. The definitions of components are also dealt with in Chapters 6 and 9.

A laser is an intense beam of light usually emitted in the infrared, visible, or ultraviolet region of the spectrum of electromagnetic radiation. The difference between ordinary light and light from a laser is that the latter is very intense, directional, monochromatic (with a single wavelength), and coherent (the waves are in step with one another). Essential elements of a laser weapon are the high-energy, laser-producing system and the beam control system, which aims the laser beam at the target and focuses the laser energy on it. Three types of lasers are considered suitable for weapon applications: a chemical laser using deuterium fluoride as the lasing medium; a gas dynamic laser using carbon dioxide; and a free electron laser (FEL). There are other types also (for details, see Chapter 6). However, a considerable amount of research is needed before such devices could be fielded. The size of such devices depends on the material used to produce the laser light, on the power output, and on whether the light is emitted in pulses or as a steady beam. Many of these questions remain unanswered.

The broad range of wavelengths at which FELs could be operated makes feasible their potential use not only as BMD and ASAT weapon, but also as communications/radar interference weapons. For example, such lasers could operate from infrared wavelengths to microwaves. In theory, at microwave wavelengths, and FEL could be used as a directional electromagnetic pulse (EMP) bomb. Such weapons are called radio frequency weapons (RFWs). The vast amount of energy released from a nuclear weapon originates from the nuclear of the nuclear explosive material. This energy is in the form of thermal radiation, blast and shock waves, and nuclear radiations consisting of gamma rays and neutrons. The gamma rays interact with air atoms and molecules to produce a sharp and relatively short but high intensity pulse of electromagnetic radiation (EMP). The latter produced in this way has no directional property. However, if a burst of radio frequency or microwave radiation were to be produced using a very high peak-power microwave generator, for example, an FEL, such an energy could be directed to a target. This type of radiation could disrupt or even destroy, for instance, electronic components on board satellites or those of communications networks and computers.

Various basing modes of these different space weapons are summarized in Table 1.1. Details of the weapons are found in Chapters 6 and 7.

Non-dedicated space weapons

A space weapon may not be aimed only at the destruction of either a satellite or a missiles in space. It could be aimed at satellite and missile command, control and space surveillance systems, which are vital to the efficient operations of spacecraft and missiles. Weapons could have ASAT capabilities but not be specifically designed as ASAT weapons. These are, for example, ICBMs and the current generation of missiles in the ABM and air defense systems. (Some of these systems are dealt with in detail in Chapters 7 and 8.) There are certain similarities in the technologies used for ATBM, BMD, and ASAT systems. Examples are target detection and tracking, guidance systems for interceptors, and, in the case of KEWs, rocket technology. Although these go a long way to fulfilling the requirement with modest changes, the ATBMs may be regarded, at least, as having residual capabilities. The overlap between various defensive system technologies and ASATs becomes more marked in the case of DEWs. (These systems are considered in detail in Chapter 8.)

An important issue to be noted here is the military capabilities of the non-dedicated systems. In the assessment of capabilities, factors such as whether they can be used or not, operational readiness, target coverage, and operational costs need to be examined (see chapter 8). It seems that the most likely application is against satellites. Thus such residual capabilities should be taken into consideration in an attempt to define space weapons.

Definition of a space weapon

The following general working definition of a space weapon is proposed (see also Chapters 6 - 9):

> A space weapon is a device stationed in outer space (including the moon and other celestial bodies) or in the earth environment designed to destroy, damage, or otherwise interfere with the normal functioning of an object or being in outer space, or a device stationed in outer space designed to destroy, damage, or otherwise interfere with the normal functioning of an object or being in the earth environment. Any other device with the inherent capability to be used as defined above will be considered as a space weapon.

A number of other definitions have been proposed in the CD.[10] However, it can be seen from the above that there are several systems that are not weapons as such but have inherent capabilities which may be converted into space weapons. Thus outer space is already potentially weaponized. However, depending on such factors as cost effectiveness, the potential weapon capabilities of some the systems may or may not be harnessed. It is hoped that the above overview might help in further discussions on the subject.

II - International Verification, an Indicator of "Peaceful" Uses of Outer Space

In defining "peaceful" uses of outer space, some technical parameters were suggested that could be adopted to indicate the nature of a spacecraft used for verification of arms control treaties. In this section, the internationalization aspect is discussed briefly (see also Chapters 3 and 5). This aspect is important because it may be a way to distinguish a satellite used for "peaceful" purposes and that used, for example, as a weapon. A "non-peaceful" application, by its nature, cannot be an international activity because it is not possible, for instance, to spread command and control processes to several nations. Thus international activities could, in general, be regarded as "peaceful."

Some of the existing treaties and current arms control measures under discussion require the so-called national technical means (NTMs) of verification. Whereas most nations still do not posses NTMs of their own, the concept of an international verification agency is becoming more acceptable, with observations from space as a critical element.

Current Debate

An international satellite monitoring agency (ISMA) to verify arms control treaties as well as to monitor crisis areas was proposed by France in 1978.[11] A UN expert group study on ISMA was published in 1981.[12] The complexity of political problems associated with the creation and operation of an ISMA led some to propose a regional satellite monitoring agency (RSMA).[13] In this context, it should be mentioned that as a result of the resolution 43/81B passed by the General Assembly of the United Nations in 1988, a Study on the Role of the United Nations in the Field of Verification has been issued by a Group of Governmental Experts.[14]

---

[10] CD document CD/851, CD/OS/WP.24, 2 August 1988.

[11] UN document A/S-10.1/7, May-June 1978.

[12] "Study on the implications of establishing an international satellite monitoring agency," UN Report A/AC 206/14, 6 August 1981.

[13] B. Jasani, "A regional satellite monitoring agency," *Environmental Conservation*, 10, No. 3 (1983), 255 - 56; and CD document CD/PV.318, 26 July 1985.

[14] UN Document A/45/372.

A number of countries have, in fact, studied the technical and financial aspects of an RSMA. For example, Canada in its PAXSAT program has studied in some depth the technical requirements to verify: (1) that outer space remains free of weapons, by monitoring activities in space from space (PAXSAT A), and (2) the reduction in conventional forces and enhance confidence-building measures in Europe, by monitoring that region from space (PAXSAT B).[15] A similar study called Project Tellus was carried out by Sweden.[16]

Monitoring activities in outer space

In June 1988, Roland Dumas, the French minister of Foreign Affairs suggested that "In view of the technical possibilities now offered to us, could we not envisage initially the constitution, within the United Nations, of an agency for the processing and interpretation of images obtained from space?"[17] He further stated that "This agency would be responsible for gathering the data obtained from civilian satellites and investigating the possible contribution that space technology could make to the implementation of multilateral programs relating to security or of a civilian nature." More recently, this proposal was reiterated in the CD by the French delegation.[18]

However, to begin with it might be better to establish international data centers where information already published could be collected and organized. This could be, for example, basic orbital information, the size, weight, and shape of satellites obtained, and information on radio frequencies used by satellites from the International Telecommunications Union. These would add to the knowledge of but not determine the mission of a spacecraft with absolute certainty. More could be learned about the earth-orbiting spacecraft by monitoring the communications flow the satellite to its command and control centers. As a next step, a system similar to the Canadian concept of PAXSAT A could be established.

Monitoring activities from outer space

As a second step toward an ISMA, to begin with, observations made routinely of the earth from outer space by civilian satellites could be collected and analyzed at the data centers. This would enable nations to learn not only how to analyze and use data from observations from outer space but also what could and could not be verified using such data. Such an activity could precede the establishment of PAXSAT B, Tellus, or other such program.

---

[15] CD document CD/PV.410, 30 April 1987; and "PAXSAT briefing graphics," Department of External Affairs, Canada, June 1987.

[16] Technical study of a verification satellite - Project Tellus, Report prepared for The Swedish Board for Space Activities, by the Swedish Space Corporation in cooperation with The Defense Staff and the Swedish Defense Research Establishment, September 1988.

[17] Address by His Excellency Roland Dumas, minister of State, minister of Foreign Affairs at the Fifteenth Special Session of the United Nations General Assembly, Thursday, 2 June, 1988.

[18] CD document CD/937, CD/OS/WP.35, 21 July 1989; and CD945, CD/OS/WP.40, 1 August 1989.

It should be noted that in August 1989, the Soviet Union proposed an idea similar to the ISMA. Under this concept, a data center under the United Nations would collect information obtained from observations made from outer space.[19] It is also proposed that the USSR would make available data from its satellites with resolutions better than 5 m. If such data were to be shared, satellites generating it would be classified as spacecraft for "peaceful purposes." The current military and civil satellites have resolutions of about 0.12 m, 10 m, respectively.

III - Summary

This brief survey indicates that whatever modus operandi is adopted in arms control negotiations (bilateral, multilateral, or both), a clear understanding of what is meant by "outer space," "peaceful" uses of outer space, and "space weapons" may well be required. Nations are rapidly developing capabilities to use outer space for scientific as well as for defense purposes. A number of treaties already exist to regulate activities of nations in this environment. However, only a limited attempt is made to define some of the relevant terms. For example, the 1972 bilateral ABM Treaty defines elements of the ABM systems. Among the multilateral treaties, only two treaties crudely define space objects. None of the agreements define the meaning of outer space or space weapons, not to mention "peaceful" uses of space. Some of these deficiencies are also repeated in new proposals.

In any discussion on the prevention of an arms race in outer space, the definition of a space weapon is critical, which in turn makes it essential to define the boundary between air and outer space. For practical reasons, it might be useful to define this as space where an object is able to complete a full orbit around the earth without further application of force. This occurs at an altitude of about 100 km. At approximately this height, a number of physical characteristics (such as temperature) of air space change also. Such a definition is called spatial definition. Another way to define outer space is to adopt the functional approach. In this case the definitions of discrete orbits, such as low earth orbits and the geostationary orbit, could be used. Thus space would be divided into various zones.

As for the meaning of "peaceful" uses of outer space, it is important that in future agreements this is either defined clearly or its use is avoided altogether. At present, nations assign different meanings to this term, which range from absence of force or conflict to non-aggressive use of outer space. The latter would permit non-aggressive military uses such as the deployment of military satellites. In the report two approaches have been suggested. One, the use of technical parameters of objects in orbits could be used to determine their nature. For example, satellites used for verification of arms control agreements could be placed in higher orbits than those used for intelligence-gathering spacecraft. Similarly, the differences in the capabilities of hardware on board civil and military satellites could be another parameter that could be used to differentiate between "peaceful" and "non-peaceful" uses of outer space. Finally, internationalization of space activities could be yet another way. An example of verification of arms control treaties is studied earlier in some detail. This was

[19] CD/OS/WP.39, 2 August 1989.

chosen because, although the verification process is a peaceful activity, it involves observations of military activities of states.

Definition of a space weapon is complicated by the fact that there are a number of other systems, non-dedicated space weapons, which could be adopted for use as a space weapon often with very minor adjustments. However, from the point of view of security, the importance of these should be considered carefully and the prospects of any arms control in this area should not be jeopardized because of the existence of the non-dedicated systems. One way may be to eliminate all categories of space weapons. An example of this generic approach would be to eliminate a whole category of space weapons. Such an approach would require clear definition of a space weapon, which would be comprehensive enough to include ASAT and BMD weapons. In this research report such a definition is proposed.

A comprehensive approach would also require definitions of such items as components and testing. In the research project five components of a space weapon are proposed: space platform, kill mechanism, energy source, surveillance, acquisition, targeting, and kill assessment system, and command, control, communications, and intelligence system. There are also several subcomponents. Examples of space tests are presented that might be considered "laboratory testing." An example is large-scale testing of C^3I systems.

The Soviet Union and the United States extensively use military satellites but by and large dedicated space weapons have yet to be deployed. As seen above, space weapons are controlled only to a limited extent. The bilateral agreements limit some BMD weapons, whereas multilateral treaties ban only the deployment of nuclear and any other weapons of mass destruction in orbits around the earth and testing of nuclear weapons in outer space. In contrast the UN Charter prohibits the threat or use of force but not the possession of weapons.

The situation may change if the two powers feel the need to deploy ASAT weapons to counter military or even civil satellites, particularly in a crisis situation or if space-based defensive BMD weapons are deployed. The continued research and development of such systems, is of particular concern. As some of the technologies for ASAT and BMD systems overlap, the former would become very effective weapons even before the latter reached a point of deployment. Thus much impetus would be given to the space weapon arms race.

The problem will no doubt become more acute when third parties, with their growing space capabilities, perceive the need to enforce their right of access to outer space with defense satellites and with ASAT weapons of their own. Thus access to outer space for non-military operations may face challenges because of technology racing ahead of legal measures. Therefore, there are arguments for enlarging discussions on arms control to include at least the other major space powers.

Bibliography

Defending Deterrence: Managing the ABM Treaty Regime into 21st Century, Antonia H. Chayes and Paul Doty (eds), (Washington: Pergamon/Brassey's, 1989).

Making Space Defense Work, A. Fenner Milton, M. Scott Davis, and John A. Parmentola (Washington: Pergamon/Brassey's, 1989).

The Pollution of Outer Space, in Particular of the Geostationary Orbit, G.C.M. Reijnen and W. de Graff (Dordrecht: Martinus Nijhoff Publishers, 1989).

The ABM Treaty: To Defend or not to Defend, W. Stutzle, B. Jasani, and R. Cowen (eds) (Oxford: Oxford University Press, 1987).

Space Weapons and International Security, B. Jasani (Oxford: Oxford University Press, 1987).

Weapons in Space, F. A. Long, D. Hafner, and J. Boutwell (eds), (New York: W.W. Norton & Company, 1986).

Anti-Satellite Weapons, Countermeasures, and Arms Control, Office of Technology Assessment, Report no. OTA-ISC-281, September 1985.

Ballistic Missile Defense Technologies, Office of Technology Assessment, Report no. OTA-ISC-254, September 1985.

Space-Strike Arms and International Security, Report of the Committee of Soviet Scientists for Peace, Against the Nuclear Threat, Moscow, October 1985.

Outer Space-Battlefield of the Future?, B. Jasani (London: Taylor & Francis, 1978).

CHAPTER 2: BOUNDARIES IN SPACE

Caesar Voûte

Arguments are cited in this Chapter in favour or against a delimitation between air space and outer space, quoting legal considerations as well as criteria based on physical properties of the different domains. Depending on the criteria used, different overlapping boundary zones are found, varying in width. It is argued that if a definition is required within the context of monitoring of compliance with a future treaty to prevent an arms race in outer space, an arbitrary but logical choice would be to define such a boundary as situated at an altitude of 100 kilometers above the earth's surface.

I - Introduction

This Chapter discusses the legal and practical reasons why it has become necessary to define boundaries in space in general, and in particular with respect to the definition of "space weapons." It summarizes the criteria that can be used to define such boundaries, taking into account the physical nature of the various layers of air space, the boundary or boundaries between air space and outer space, and the various zones that can be distinguished in outer space such as LEO, the GSO, and deep space. The relevance of any such boundaries with respect to the definition of "arms in space" is noted. In doing so, use is made of the debate from 1959 onward in the UN Committee on the Peaceful Uses of Outer Space (UNCOPUOS) and in its Legal and Scientific-Technical Sub-Committees, and of publications by a number of experts on international law and on space law.

It is concluded that in any treaty for the prevention of an arms race in outer space, an arbitrary limit should be adopted for pragmatic reasons as the boundary between air space and outer space. Such a boundary, for instance at an altitude of 100 km, should be applicable only in the context of the treaty and should be easy to monitor.

II - Some Legal Considerations

Even though the question of the boundary between air space and outer space has been on the agenda of the UNCOPUOS and its Legal and Scientific-Technical Sub-Committees, since 1959, there have been no concrete results regarding a demarcation of outer space. For this reason The Netherlands and several other Western countries proposed at the 31st UNCOPUOS session in June 1988 to remove the term "the boundary between air space and outer space" from the agenda of the Legal Sub-Committee.[1] The discussion in UNCOPUOS, it was reasoned, could always be reopened as and when necessary.

The demarcation of outer space has also been a subject of considerable debate among experts on international law and on space law, resulting in an impressive number of publications on the subject.[2]

Understandably, this book cannot discuss the uses of space without addressing the problem of how the boundary between air space and outer space should be defined and demarcated. Yet, it would be impractical -- and also fundamentally incorrect -- if a book on the subject, published within a UN context, would attempt to define the boundary between air space and outer space independently and without making reference to the work done by UNCOPUOS and its Legal and Scientific-Technical Sub-Committees.

However, it is also well-known that some countries are not in favour of a physical demarcation of outer space. Several reasons are quoted to substantiate this opposition. It may in practice lead to an extension of the area over which sovereignty is being claimed. It could result in disagreements about the right of passage of spacecraft through national air space on their way to and from outer space and on the status of spacecraft during various parts of their flights, etc., without the need for a precise demarcation between air space and outer space having been convincingly demonstrated.

It has also been pointed out that the motion of aircraft and spacecraft is governed by different physical principles: support from the atmospheric air in the case of aircraft; absence of any such support and even of frictional resistance ("air drag") in the case of spacecraft. Different environmental conditions are required for these principles to operate: a sufficient high air density in the case of aircraft motion; a sufficient low density in the case of spacecraft motion.

At present (and for the near future), the domains of space where these environmental conditions are met do not overlap, or even touch, although it is admittedly difficult to define a sharp, accurately measurable boundary where one domain ends and the other begins. In practice aircraft (and balloons) remain at altitudes well below 60 km, whereas no orbital space

---

[1] "Report of the Committee on the Peaceful Uses of Outer Space," UN Document A/43/20, 19 July 1988.

[2] M. Benkö, and W. de Graaff, "Questions relating to the definition/delimitation of outer space and outer activities and the character and utilization of the geostationary orbit," in M. Benkö, W. de Graaff, and G.C.M. Reijnen, *Space Law in the United Nations*, (Dordrecht: Martinus Nijhoff, 1985), pp. 121 - 44; St. Frhr. von Welck, and R. Platzöder, (eds): Weltraumrecht -- Law of Outer Space, Textsammlung / Legal Documents, (Baden-Baden, FRG: 1987) Nomos Verlagsgesellschaft, R. Wolfrum, "Rechtliche Ordnung des Weltraums", in K. Kaiser and St. Frhr. von Welck, (eds): *Weltraum und internationale Politik*, Schriften des Forschungsinstituts der Deutschen Gesellschaft für Auswärtige Politik e.V. (Bonn: R. Oldenbourg Verlag München, 1987), pp. 142 - 252; "The question of the definition and/or delimitation of outer space," UN Document A/AC.105/C.2/7, 7 May 1970; and "The question of the definition and/or delimitation of outer space," UN Document A/AC.105/C.2/7 Add. 1, 21 January 1977.

flight is possible, even at perigee (the orbit point nearest the earth), at altitudes lower than 90 and 100 km.

The problem is complicated by the fact that air density, which determines the possibility of aircraft flights below 60 km and space flights above 100 km altitude, not only varies with altitude, but, at a given altitude, also with time. Moreover, some space vehicles operates during one part of their mission as a spacecraft and during another part as an aircraft. The U.S. space shuttle, the USSR Buran spacecraft, and the European Space Agency (ESA) Hermes spacecraft now under development are examples of this category. In contrast, even in this case the transition from one environment to the other is accompanied by violent interactions with the environment, such as heating of the vehicle due to air friction and loss of radio communications with the outside world, occurring generally between altitudes of 100 and 50 km.[3]

Proponents of a definition/delimitation argue that the basic difference between the legal status of air space and outer space necessitates such a delimitation from a legal point of view. Otherwise, the legal system would remain incomplete and in some respects ambivalent.

The leading argument against the necessity of such a delimitation points out that all legal problems that have arisen until the present have been solved without an agreed solution to the delimitation question. The reason is the fact that, for example, in case of accidents, the application of the relevant provisions of the 1967 Outer Space Treaty and of the 1972 Liability Convention do not depend on the location *where* the accident happened, namely, in air space or in outer space. It is only relevant that the damage has been caused by a *space object.*

It is sometimes also said that any precise delimitation of outer space agreed upon at present might become obsolete in the future as a result of new technological developments. However, it is very difficult to anticipate exactly what would be the consequences of such developments for the present delimitation problem. Therefore for the time being, in the opinion of many it seems better to proceed from the present situation and its expected changes in the near future and to base any legal consideration on these.

Under such circumstances, one can argue why it is necessary or even unavoidable in the case of a Treaty on the prevention of an arms race in outer space to define a precise boundary between air space and outer space. Furthermore, it must be concluded that if in such a treaty a precise boundary is to be adopted, for pragmatic reasons, as the boundary between air space and outer space, such a boundary would be arbitrary by nature, should be easy to monitor, and should be applicable only in the context of this particular treaty. However, the adoption of such an arbitrary boundary within the context of a specific treaty should not be construed or accepted as a precedence as "*the* boundary between air space and outer space" in the sense of international space law. It need not be more generally applicable

[3] Benkö and de Graaff, "Questions," *op. cit.*

in other cases where the delimitation between air space and Outer Space is under consideration.

III - Need to Define the Boundary between Air Space and Outer Space

We argue here for the necessity of a proper demarcation of the boundary between air space and outer space because it could contribute essential elements for the elaboration and negotiation of a treaty on the prevention of an arms race in outer space. Without such arguments, one might otherwise better leave the boundary question open, simply referring to the debate in UNCOPUOS.[4]

Space weapons can be divided according to deployment mode and with respect to their targets (see Table 1.1 in Chapter I) into:

. Ground-to-space.
. Space-to-space.
. Space-to-air.
. Space-to-ground.
. Air-to-space.

We are dealing therefore with weapons either based in space (including pop-up systems, aimed at targets in space), in the air, or on the ground (on land or in the sea), or with weapons aimed at targets in space, based on the ground (on land or in the sea) or in the air. Within the context of a treaty for the prevention of an arms race in outer space, there must be clear distinctions between weapons based in the air and weapons based in space, and between weapons aimed at targets in space and targets based in the air.

[4] "Rapport du sous-comité juridique sur les travaux de sa vingt-quatrième session (18 mars - 4 avril 1985)," UN Document A/AC/105/352, 11 April 1985; "Report of the Committee on the Peaceful Uses of Outer Space," UN Document A/40/20, 26 July 1985; "Rapport du sous-comité juridique sur les travaux de sa vingt-cinquième session (24 mars - 11 avril 1986)", UN Document A/AC.105/370, 5 May 1986; "Report of the Committee on the Peaceful Uses of Outer Space," UN Document A/41/20, 26 June 1986; "Report of the Legal Sub-Committee on the work of its twenty sixth-session (16 March - 3 April 1987)," UN Document A/AC.105/385, 16 April 1987; "Report of the Scientific and Technical Sub-Committee on the work of its twenty-fifth session," UN Document A/AC.105/409, 1 March 1988; "Outer space Sub-Committee continues general exchange of views, debate on definition of outer space," UN Information Service, Geneva, press release OS/330, 24 March 1988; "Report of the Legal Sub-Committee on the work of its twenty seventh session (14 March - 31 March 1988)," UN Document A/AC.105/411, 8 April 1988; "Report of the Committee on the Peaceful Uses of Outer Space," UN Document A/43/20, 19 July 1988; and St. Frhr. von Welck, and R. Platzöder, *Weltraumrecht, Law of Outer Space*, Textsammlung / Legal Documents, (Baden-Baden, FRG: 1987) Nomos Verlagsgesellschaft.

Monitoring of compliance with a treaty prohibiting space weapons, therefore, requires an operational definition of the boundary between air space (where certain activities are allowed) and outer space (where comparable activities are banned).

Furthermore, testing of any nuclear explosive and the deployment of certain weapon systems, space-based nuclear weapons, X-ray lasers, and non-nuclear space-to-air or space-to-ground lasers -- are already banned by treaties or customary international law (see Chapters 1 and 3). This is in spite of the absence of a treaty on the prevention of an arms race in outer space. Hence it is already important to delimitate now air space (where deployment of certain weapons would still be authorized in principle under present-day conditions) from outer space (where deployment of some weapons is not authorized). Details of permitted and prohibited activities in outer space are discussed in Chapter 3.

With an adequate definition, we could determine which activities are presently prohibited in the absence of a treaty, and which are still allowed by existing treaties and/or by customary law. Outer space in this context includes LEO, the GSO, and deep space.

Similarly it would be possible to assess all defined testing programs, relating them to prohibited or allowed activities, once again on the basis of a proper delimitation between air space and outer space. In the case of a general ban of ASAT weapons it would also be important to accurately define the boundary between air space and outer space to determine when a potential target for such weapon systems can be deemed a "satellite" deployed in outer space, or when the potential target is limited to the air space.

IV - Criteria for the Definition of Boundaries in Space

Here we attempt to provide some precise criteria for the demarcation between air space and outer space relating to specific aspects of an arms race in outer space. The identification and definition of these criteria take into account the discussions in Chapters 4, 6, and 7.

It appears less productive to analyze the various positions of states with regard to proposals submitted to the UNCOPUOS Legal Sub-Committee because they are general in nature and not directly related to preventing an arms race in outer space. Also, the debates in UNCOPUOS and its Legal Sub-Committee have tended in recent years to put more emphasis on the character and utilization of the GSO (geostationary orbit), which constitutes another boundary in space, than on the exact boundary between air space and outer space.

As a solution to the delimitation question, two approaches are under consideration in the legal field, the spatial and the functional approach. Each involves criteria that for our purposes here, need to be correlated with the technical parameters of space weapons.

The Spatial Approach

The spatial approach tries to establish the lower boundary of outer space.[5] As to the altitude, no less than eight possible criteria have been identified. They are demarcation based on:

1. The equation of the upper limit of national sovereignty with the concept of "atmosphere".
2. The division of atmosphere into layers.
3. The maximum altitude of aircraft flight (theory of navigable air space).
4. Aerodynamic characteristics of flight instrumentalities (von Karman line).
5. The lowest perigee of an orbiting satellite.
6. The earth's gravitational effects.
7. Effective control.
8. The division of space into zones.

The division of the atmosphere into layers takes into account a number of physical parameters that characterize each. However, one should note that boundaries thus defined are not sharp; one observes some scattering of measurements above and below them. Furthermore, such boundaries vary with time, for example, in a day-and-night cycle, or seasonally.

Among the physical parameters, temperature must be mentioned because it tends to define four layers reasonably sharply (Fig. 2.1). Within the *troposphere*, the lowest earth layer extending to an altitude of about 20 km, the temperature falls to about -85° C. Above 20 km, the temperature rises, marking a boundary between the troposphere and the *stratosphere*.[6] The temperature in the stratosphere increases to about -2.5° C until an altitude of about 50 km. The warming of the stratosphere is due to the presence of ozone. The ozone layer absorbs much of the solar ultraviolet radiation, causing considerable heating in this layer. Beyond 50 km, the temperature starts to decrease again, defining a boundary between the stratosphere and the third layer, the *mesosphere*.

The temperature in the mesosphere, which extends between 50 and 85 km, falls to about -135° C. Beyond 85 km, the temperature rises with altitude, thus separating mesosphere from *thermosphere*. Heating of the thermosphere is due to absorption of highly energetic solar radiation by the small amount of residual molecular oxygen that is present. Farther out into space the atmospheric density becomes even less and the temperature rises even more purely because of the solar radiations. The temperature minimum between the mesosphere and the thermosphere is variable in time as a function of solar exposition and solar activity.

---

[5] See publications listed in note 2.

[6] "Upper Atmosphere Research Satellite," NASA brochure, Washington, DC, 1989.

Fig. 2.1 Variation of Atmospheric Temperature with Altitude.

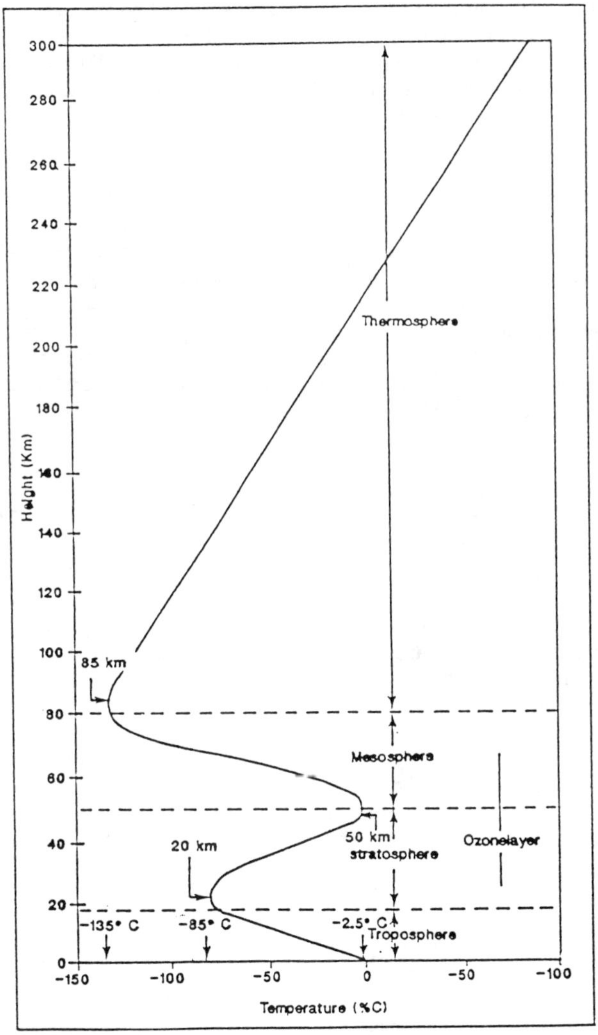

Source: Upper Atmosphere Research Satellite, NASA brochure, 1989.

The competition of the mixture of gases in the atmosphere provides another criterium. Troposphere, stratosphere, and mesosphere together represent the *heterosphere*, where because of a sufficient mixture the same proportion of gases is maintained as on the earth's surface. Therefore, one always finds "air" (as a mixture of nitrogen and oxygen) up to an altitude of about 90 - 100 km. This can be considered as the air space. Above that altitude, in the so-called *homosphere*, the gases (oxygen and nitrogen) change their relative concentrations according to individual molecular weight. This is due to the fact that the low atmospheric density no longer allows for a sufficient mixture because of the absence of a sufficient number of collisions between the various gas molecules. This region can be considered to be outer space.

The demarcation based on the maximum altitude of aircraft flight (theory of navigable air space) and the demarcation on aerodynamic characteristics of flight instrumentalities (von Karman line) take into account the minimum airlift needed, at whatever speed can be attained, to maintain the altitude of an airplane using the power of air-breathing engines and to permit its maneuverability with the help of a rudder and/or wingflaps. This demarcation is based on the minimum air pressure required. The maximum altitude where an aircraft as such can operate is at present well below 60 km above sea level. This boundary corresponds more or less to the boundary between stratosphere and mesosphere, where a temperature reversal occurs at an altitude of about 50 km.

The demarcation of the lowest perigee of an orbiting satellite is defined by the region in which it can still describe a full orbit around the earth according to the laws of celestial mechanics without any additional propulsive force. At altitudes *below* 100 - 110 km, this is impossible, because due to the air drag of the residual atmospheric density, the satellite would burn up or re-enter the earth's atmosphere. However, this minimum altitude where a satellite can still describe a full orbit around the earth depends significantly on the eccentricity of the orbit, on the area-to-mass ration of the spacecraft, and on the density profile of the atmosphere (which changes with time). This boundary of about 100 km lies only slightly beyond the 85 km zone at which the final temperature reversal occurs at the boundary between the mesosphere and thermosphere.

Tethered satellites are an important exception to the rule that a satellite can describe a full orbit around the earth only above an altitude of 100 - 110 km. For tethered satellites, the lower satellite can move well below an altitude of 100 - 110 km provided that it is maintained in its orbit by the upper satellite moving sufficiently above this altitude.

When discussing a definition of the boundary between air space and outer space using as a criterium the lowest altitude at which a satellite can still describe a full orbit around the earth, one must also keep in mind proposals to use the atmosphere of the earth to decrease the energy of interplanetary spacecraft.

Some weapon systems can function only in an outer space environment, for example, neutral particle beams, because of the absorption of the particle beam in the atmosphere. In practice such space weapons could still be effective in a transition zone with sufficiently thin air between altitudes of 70 and 110 k. Here the demarcation of air space from outer space is also a matter of convenience, irrespective of whether the weapon system is allowed or prohibited under existing treaties and customary law. Thus it can be seen that each atmospheric region can be defined by characteristic ranges of temperatures, air density and composition. Moreover, as we see later, some of the temperature reversals occur more or less at the maximum altitude of aircraft flight or near the lowest altitude where a spacecraft can still describe a full orbit around the earth.

There were also theories based on combinations of the various approaches. For several years, however, only the theory on the demarcation according to the lowest perigee of an orbiting satellite has been advocated by those states that favour a spatial delimitation. The other delimitation theories are no longer under discussion.[7]

The following arguments have been expressed in favour of a delimitation according to the lowest perigee of an orbiting satellite at an altitude of approximately 100 km above sea level without taking into account the special case of tethered satellites.

1. It would satisfy the demands of free space flight and space exploration, for it defines outer space as the region in which a satellite can still describe a full orbit around the earth. At altitudes *below* 100 - 110 km, this is impossible, because air drag would cause the satellite to burn up or re-enter the earth's atmosphere.
2. The needs of international civil aviation would be taken care of, because this delimitation lies well above the maximum altitude where an aircraft as such can operate, which is at present well below 60 km above sea level. Thus, as a rule, interferences between air traffic and outer space activities, as well as between the different legal regimes that are applicable to air space and outer space can be avoided.
3. This theory would be in agreement with actual practice, because so far no state has protested against satellites passing over its territory at altitudes above 100 - 110 km (under regular, *i.e.*, non-accident, conditions).

A convenient definition could therefore be:

> Outer space is all of the space surrounding the earth where objects can move in at least one full orbit without artificial propulsion systems according to the laws of celestial mechanics, without being prevented from doing so by the frictional resistance of the earth's atmosphere. It extends from an altitude of approximately 100 km upwards.[8]

---

[7] For details of those theories, see the two relevant UNCOPUOS studies in note 2.

[8] G.C.M. Reijnen, and W. de Graaff, *The Pollution of Outer Space, in Particular of the Geostationary Orbit*, Utrecht Studies in Air and Space Law, (Dordrecht: Martinus Nijhoff Publishers, 1989), p. 3.

The functional approach

According to the functional approach to the delimitation question, the legal regime governing outer space should be based primarily on the *nature* and *type* of particular space activities. A distinction is made between aeronautical and astronautical activities, wherein the latter should be subject to space law, irrespective of the altitude at which they are carried out.

Such an approach would render it unnecessary to solve the theoretical dispute of whether there is a boundary between air space and outer space and where it should be located. Instead, however, the functional solution raises problems concerning the definition of the terms "space activity" and "space flight" as well as the distinction between "aircraft" and "spacecraft." For example, a consequence of the functional approach might be that a spacecraft would be exclusively subject to the rules of space law from launch to landing, even when it crosses the sovereign air space of a foreign state.

It is argued that the functional approach would permit a definition of objectives and missions for space activities and that regulations could be established to avoid possible interferences between *peaceful* activities and activities that could have adverse consequences for human life on earth. However, a substantial proposal, offering a functional solution to the problem of delimitation has not been submitted to legal sub-committee during the last few years.

Nevertheless, the functional approach must not be neglected because of its practical value for the allocation of suitable and specific satellite orbits for various civilian and military applications.[9] This is the more important since the number of available slots on these orbits is limited. With the growing frequency of near-earth space activities, this is gradually creating a number of problems, calling for rules-of-the-road measures.[10]

Figure 2.2 presents an overview of usable satellite orbits around the earth for a variety of uses: geostationary weather satellites, TV and communication satellites, satellite-based navigation systems, earth-observation satellite functioning in various parts of the electromagnetic spectrum (visible, near infrared and far infrared, and microwave bands). In some cases it is possible to use polar and equatorial orbits as well as orbits under any angle with the earth axis, whereas the equatorial orbit is the only one possible in other cases like the geostationary orbit.

---

[9] C. Voûte, "Economische perspectieven van de ruimtevaart," in *Los van de Aarde*, van Niekerk, W., (ed.): (Meppel, Netherlands: de Feniks Pers, Meppel, Netherlands, 1987), pp. 35 - 55.

[10] H. Feigl, "Objectives, Realisable Possibilities and Problems of a Multilateral Protection Regime for Outer Space," Conference on Disarmament, Ad hoc Committee on the Prevention of an Arms Race in Outer Space. Statement made on behalf of the delegation of the Federal Republic of Germany, Geneva, 11 July 1989.

Fig. 2.2 Usable satellite orbits around the earth.

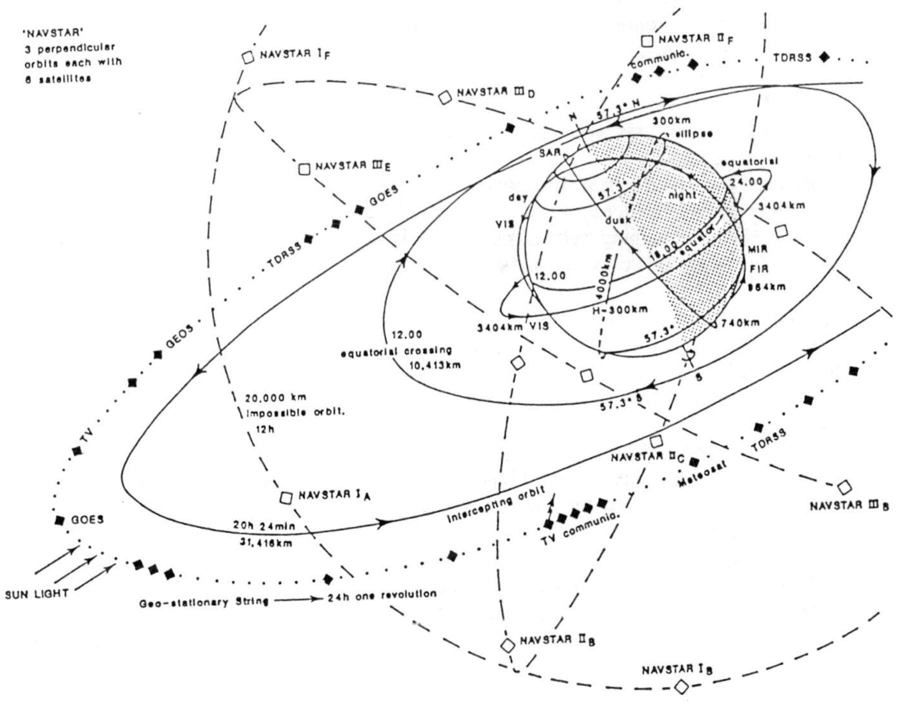

Source: W. van Niekerk, et al. "Los van de Aarde." (Meppel: De Feniks Pers, 1987), p. 40.

As can be seen from Figure 2.2, LEO are available at discrete altitudes between 100 km and 10,413 km, the actual altitudes selected depending on the type of applications, the possible damaging effects to spacecraft and payload encountered in the radiation belts, the number of revolutions around the earth, the revisit cycle of any point on earth, and the planned duration of the missions. They are followed by a second zone at an altitude of 20,000 km with a 12-hour revisit cycle for the NAVSTAR satellite-based navigation system.

A very important zone is found at an altitude of about 36,000 km (or more precisely 35,786.56 km above the earth's equator) where the revolution of a satellite around the earth takes one day (23 hours, 56 minutes, 4 seconds). If positioned in an equatorial orbit, satellites moving in the same direction as the earth rotation with a velocity of 3,074.7 m/s remain in the same position with respect to a particular point on the equator, permanently viewing about one-third to one-quarter of the earth's surface.

These fixed positions on the GSO are of particular value for meteorological satellites programmed to make very frequent observations (at 1-hour to 15-minute intervals), and for communication and direct broadcasting satellites, which maintain permanent communications with ground receiving facilities within their field of view. Tracking and data relay satellite systems (TDRSS) maintaining permanent communications with many other space missions are also positioned on this orbit for the same reason.

If one visualizes the GSO as a single narrow tube in the equatorial plane at an altitude of about 36,000 km above the earth's surface, it is easy to understand that the number of positions ("slots") on this orbit is limited. Already now in some orbital sections, the smallest distances between satellites are no more than 20 km.[11]

Satellites cannot be positioned too near to one another on this orbit because of the risk of electromagnetic or even physical (collision) interference. Hence GSO constitutes another highly valuable but limited natural resource, the excessive use of which can give rise to sensitive political issues.

Beyond the altitude of 36,000 km, we come gradually into the realm of deep space where the trajectories of satellites or their debris are less and less influenced by earth's gravity. The effects of air drag, the braking force of frictional resistance of the residual air, will already decrease considerable beyond an altitude of 1,000 km.

Even at some 1,000 km altitude, this friction is still present, albeit very weak, and it will take an object at this altitude at least some 1,000 years to descend to the earth's surface. Burned-out satellites and other space debris moved into this region will normally stay in space for thousands of years before moving slowly to lower positions from which they can eventually fall back on earth. Hence the present policy to use this region as a storage area ("dust bin") for burned-out satellites. Nevertheless, the region of LEO as defined above is

[11] Ibid.

becoming more and more crowded with manufactured space debris gradually constituting a grave risk for civilian and military space operations.[12]

The main value of deep space is its use for the exploration of the moon and other celestial bodies and for astronomical studies because of the unique possibilities to monitor and measure various physical processes without interference of the near-earth environment, in particular the atmosphere.

V - Relevance of Demarcation Criteria

In this section the criteria of the spatial and functional approaches are compared with the legal acceptability and the technical parameters of space-, air- or ground-based ASAT, ATBM, and BMD weapons; types of space-based systems that might be used against ground facilities; the nature of interference with the normal functioning of satellites in earth orbits and characteristics of such systems; components and sub-components of space weapons (for example, a ground-based laser with a mirror, a component, in space); and the boundaries in space from the perspective of the definition of testing.

Space weapons can be subdivided according to deployment mode into (see Table 1.1):

- Ground-space.
- Space-space.
- Space-air.
- Space-ground.
- Air-space.

We are dealing therefore with weapons either based in space aimed at targets in space, in the air, or on the ground, or with weapons aimed at targets in space, based on the ground or in the sea. These weapons can be further subdivided into nuclear and non-nuclear weapons (see Table 1.1).

The deployment of some of these systems is banned by a number of treaties or Customary International Law (see Chapters 1 and 3). It is therefore important to delimitate air space, where their deployment would still be authorized in principle unless prohibited by other treaties, from outer space, where their deployment is not authorized. Also testing of some of these devices, such as nuclear explosives, and any hostile act are prohibited in outer space or in air. Moreover, any physical interference, whether or not resulting in damage, with space assets of another country without that country's authorization (e.g., unauthorized inspection of another state's satellite, is also prohibited.

[12] H.A. Baker, *Space Debris: Legal and Policy Implications*, Utrecht Studies in Air and Space Law, (Dordrecht: Martinus Nijhoff, 1989); G.B. Field, N.J. Rees and D.N. Spergerl, "Is the space environment at risk," *Nature*, 336, No. 6201 (1989), 725 - 26; Reijnen and de Graaff, "Pollution of Outer Space."

We are dealing here with two issues, that of weapon systems deployed in space targeted at objects in the air or on the ground, and that of weapon systems deployed on the ground, in the air, or in space targeted at objects in space. This indicates the need to know whether a weapon system and a target are deployed in air space or outer space. Again, it is essential to have available an adequate definition of the boundary between air space and outer space in order to determine which activities are prohibited and which ones are basically allowed by existing treaties and/or by customary law. Outer space in this context includes all outer space: LEO, the GSO, and deep space.

If the spatial approach is adopted for the definition of this boundary, the most suitable definition appears to be that based on the lowest perigee of an orbiting satellite, which is applicable to targets in space and to most weapons deployed in space. The definitions based on the determination of atmospheric layers as a function of temperature, air density, or atmospheric composition do not take into account the characteristics of space weapons.

Nevertheless such a definition would not cover all cases. In the first place, tethered satellites are an exception to the rule of lowest perigee of an orbiting satellite noted earlier. Furthermore there are space weapons such as neutral particle beams, which are still effective in the upper part of air space as defined in this manner.

In the case of a general ban of ASAT weapons, the spatial delimitation of air space from outer space applies as well in order to determine when a potential target for such weapon systems can be deemed to be a "satellite" deployed in outer space, or when the potential target is limited to the air space.

In all cases the spatial approach toward the delimitation of air space from outer space applies, either because of its legal context or in view of the physical nature of the boundary concerned. Nevertheless, none of the criteria discussed earlier permits us to define a very sharp boundary and none of these boundaries is applicable under all circumstances. There are always transition zones, differing sometimes in position and width depending upon the particular space-related missions or activities (Fig. 2.3). Moreover, boundary zones may vary in time, adding to the difficulty of monitoring treaty compliance. For monitoring to be effective and unequivocal, it should be based on a sharp boundary easy to observe.

This means in practice that in various cases an *arbitrary* boundary will have to be adopted for the purpose of monitoring compliance to a particular treaty. It appears logical to draw such an arbitrary boundary as close as possible to the boundaries defined by measurable physical parameters. In practice this signifies that an arbitrary boundary between air space and outer space at an altitude of 100 km would be the logical choice.

In a similar way it is possible to assess all defined testing programs or parts of programs, relating them to prohibited or allowed activities as specified in the preceding paragraphs, once again on the basis of a proper delimitation between air space and outer space.

Fig. 2.3 Boundaries between air space and outer space based on a spatial approach.

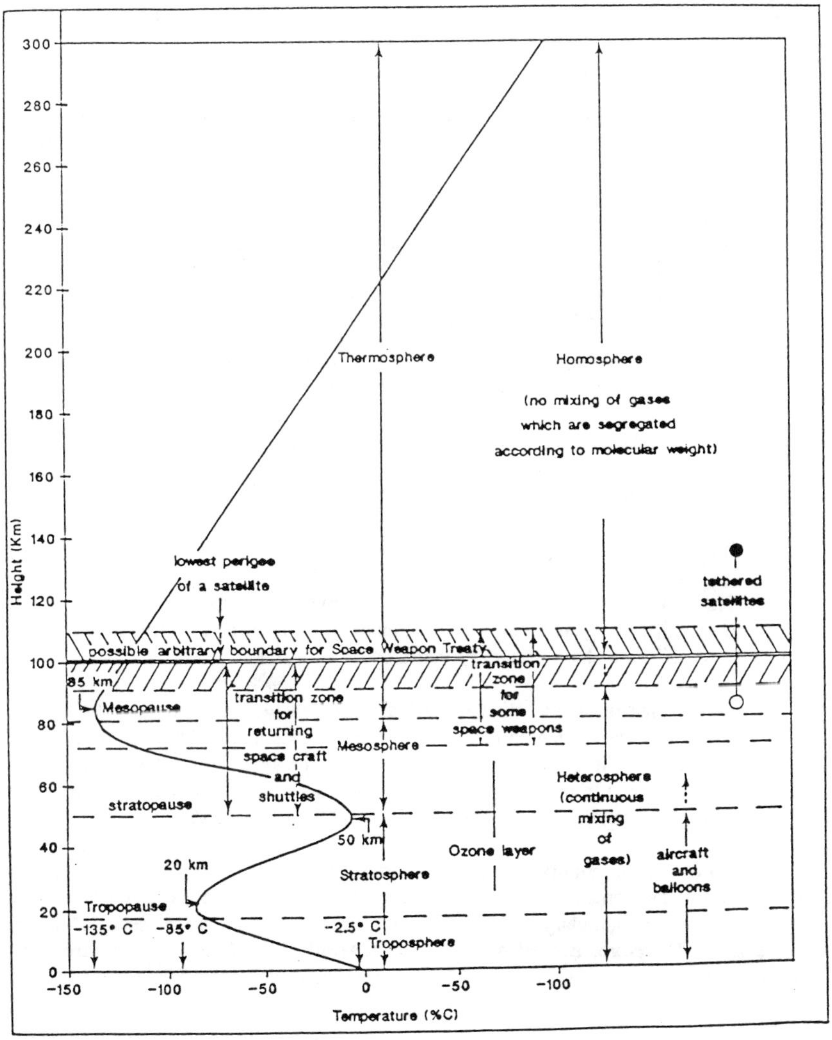

The boundaries in space from the perspective of the definition of testing should be defined either by the special approach or by the functional approach depending upon the circumstances of the particular testing program. We are dealing with a system that can be used in two different modes: either in an ASAT or ABM mode (as a "space weapon") or in an air defense mode, or with a system that is suitable only to operate in a single mode subject to the considerations covering space weapons as described earlier.

VI - Conclusions

In the discussions on a treaty on the prevention of an arms race in outer space, it is important to define space weapons that in terms makes it essential to define, as accurately as possible, the boundary between air space and outer space. The latter includes LEO, GSO, deep space, and the celestial bodies. A suitable definition might be:

> Outer space is all of the space surrounding the earth where objects can move in at least one full orbit around the earth without artificial propulsion systems according to the laws of celestial mechanics, without being prevented from doing so by the frictional resistance of the earth's atmosphere. It extends from an altitude above the earth of approximately 100 km upward.

It should be noted that some spacecraft (tethered satellites) and some space weapons (such as neutral particle beams) do not fit into this definition.

This boundary between air space and outer space corresponds more or less with the boundary between the mesosphere and the thermosphere at an altitude of 85 km above the earth's surface characterized by a temperature reversal. It also corresponds more or less with the boundary between heterosphere and homosphere at an altitude of 90 - 100 km where the composition of the atmospheric gases changes.

The various boundary zones may vary in time, adding to the difficulty of monitoring treaty compliance using such definitions. For monitoring to be effective and unequivocal it should be based on a sharp boundary that is easy to observe.

This means in practice that in various cases an *arbitrary* boundary will have to be adopted for the purpose of monitoring compliance to a particular treaty. It appears logical to draw such an arbitrary as close as possible to the boundaries defined by measurable physical parameters. In practice this signifies that an arbitrary boundary between air space and outer space at an altitude of 100 km would be logical choice.

However, the adoption of such an arbitrary boundary between air space and outer space within the context of a specific treaty should not be construed or accepted as a precedence as "*the* boundary between air space and outer space" in the sense of international space law. Therefore, it need not be more generally applicable in other cases where the delimitation between air space and outer space is under consideration.

This spatial approach to the boundary between air space and outer space applies to all situations where such a delimitation is required in relation to the prevention of an arms race in outer space. In a few cases the functional approach is applicable simultaneously, based on the nature and the functions of the space activities concerned. However, from a political and legal point of view the spatial approach should have priority.

The functional approach is mainly valid if one wishes to restrict the military use of LEO and GSO, which constitute limited natural resources of the earth's environment in order to protect the interest of civilian uses of outer space. In this case protective measures against excessive pollution of these regions of outer space by debris produces by military launches and decaying military spacecraft are to be included.

CHAPTER 3: THE LEGAL ASPECTS OF PEACEFUL AND NON-PEACEFUL USES OF OUTER SPACE

Ivan A. Vlasic

This Chapter examines the origin and evolution of the term "peaceful uses of outer space" as employed in official government pronouncements and multilateral treaties, against the background of state practice. The examination leads to the conclusion that this term, still without an authoritative definition, has been a source of considerable confusion and controversy and its use should therefore be avoided in future limitation agreements. However, if employed, it should in each such agreement be defined with great precision. In this Chapter, all military uses of outer space, both current and those in the process of development, are placed in three broad categories: (1) those prohibited by international law, (2) those that are lawful, and (3) those that should be banned because of their destabilizing effect on international peace and security.

The UN General Assembly Resolution 42/68 on International cooperation in the peaceful uses of outer space, adopted without vote on December 2, 1987, reminds in part:

> *Deeply convinced* of the common interest of mankind in promoting the exploration of mankind in promoting the exploration and use of outer space for peaceful purposes ...
>
> *Gravely concerned at* the extension of an arms race into outer space,
>
> *Urges* all States, in particular those with major space capabilities, to contribute actively to the goal of preventing an arms race in outer space as an essential condition for the promotion of international co-operation in the exploration and use of outer space for peaceful purposes, [and]
>
> *Requests* the Committee on the Peaceful Uses of Outer Space to continue to consider, as a matter of priority, ways and means of maintaining outer space for peaceful purposes ...

The adjective "peaceful" in relation to outer space activities can be found in virtually all UN documents devoted to outer space matters as well as in the space law treaties. Yet, more than three decades since the launching of Sputnik I, the term "peaceful" as employed in the context of international space law remains without an authoritative definition. The widely accepted interpretation given this key term of space law prior to and immediately after the advent of the space age, namely that "peaceful" means "non-military," was soon contradicted by the practice of states, primarily the United States and the Soviet Union. In the period between October 1957 and the adoption of the Outer Space Treaty (OST) in January 1967, these two space powers had placed into orbit a number of military payloads and had come increasingly to rely on space technology in their military planning.

A major purpose of this Chapter is to demonstrate the difficulties encountered by governments and analysts in the interpretation of the terms "peaceful uses" and "peaceful purposes." These terms remain the object of contradictory interpretations even though the principal space powers had tacitly agreed some time ago to treat all military activities in outer space as permitted except those explicitly prohibited by treaty or customary law. An examination of various multilateral agreements that employ the phrase "peaceful purposes" tends to support the conclusion that "peaceful" was meant to mean, if not always the opposite of "military", then certainly the opposite of "testing and deployment" of weapons, the only clear exception being the 1982 United Nations Convention on the Law of the Sea. Yet such an anti-military interpretation runs counter to the practice of states, especially those with significant space capabilities. Even the term "military" in the context of space activities and space uses creates problems in interpretation because civilian satellites, such as Landsat, SPOT, Gorizont, and Molniya, can be and are being used for military purposes. It is therefore recommended that the adjective "peaceful" be avoided in future outer space and disarmament agreements; instead, each particular agreement should clearly and precisely identify activities, uses or devices that are to be prohibited. This Chapter offers a catalogue of activities and uses of outer space that are prohibited under current international law an, for purposes of illustration, identifies those uses of outer space that should be banned because of their detrimental effect on international peace and security.

Since the term "peaceful" continues to be used both in the UN documentation on outer space as well as in the practice of states, it should be of interest briefly to examine the origin of this seemingly elusive and ambiguous adjective that shows remarkable vitality.

I - "Peaceful" -- Its Origin and Use in the Early Years

The term "peaceful" in connection with outer space uses made its appearance even before the first artificial satellite was launched into orbit. Thus in January 1957, Ambassador John Cabot Lodge, addressing the UN General Assembly on behalf of the United States, expressed the hope of his government that "future developments in outer space would be devoted exclusively to peaceful and scientific purposes."[1] To achieve this objective. Lodge proposed that the testing of satellites and missiles be placed under "international inspection and participation."[2]

A few months later, Secretary of State John Foster Dulles announced the willingness of the United States to devise "a system which would insure that outer space missiles would be used exclusively for peaceful and scientific purposes" and "for the benefit of mankind."[3] In August 1957, four Western powers -- Canada, France, the United Kingdom, and the United States -- in a joint submission to the UN Disarmament Commission called for a study of an inspection system that would assure that the launching of objects through outer space would be "exclusively for peaceful and scientific purposes."[4] On November 14, 1957, the UN

[1] Quoted in P. Jessup and H. Taubenfeld, *Controls for Outer Space and the Antarctic Analogy* (New York, Columbia University Press, 1959), p. 252.

[2] Ibid.

[3] Ibid., p. 253.

[4] United Nations, *the United Nations and Disarmament 1945 - 1970* (UN Publication No. 70.IX.1, 1970), p. 67.

General Assembly adopted (by 56 votes to 9, with 15 abstentions) resolution 1148 (XII), which incorporated almost verbatim the four power proposal. This was a landmark document, not only because it represented the first General Assembly resolution on outer space but also because it introduced the phrase "exclusively for peaceful purposes" in an authoritative UN text.

President Dwight Eisenhower also took part in the debate. In January 1958, in a letter to Soviet Premier N. Bulganin, Eisenhower proposed that the two nations agree to use outer space "only for peaceful purposes" and not for the "testing of missiles designed for military purposes."[5] This statement, together with similar pronouncements made earlier by high U.S. officials, suggests that the term "peaceful," even after the launching of Sputnik I, was being used by the Untied States in contradistinction to "military" in the context of outer space activities.

The 13th session of the UN General Assembly, held near the end of 1958, provided a forum for the first wide-ranging multinational debate on "Questions of the Peaceful Use of Outer Space." In the debate, virtually all participants used the term "peaceful" as an antonym to "military," with the representative of Sweden echoing the common sentiment with his appeal to UN member states to "safeguard outer space against any military use whatever."[6] Following an extensive discussion, largely centered on the Soviet proposal to ban both the use of outer space for military purposes and military bases on foreign soil, the General Assembly adopted resolution 1348 (XIII), which recognized as the "common aim" of humankind that outer space "should be used for peaceful purposes only." One important result of this resolution was that it established the ad hoc Committee on the Peaceful Uses of Outer Space (COPUOS), which became a year later, by resolution 1472 (XIV) of December 12, 1959, a regular committee of the General Assembly. However, neither of these resolutions, nor the comprehensive report of the ad hoc Committee released on July 14, 1959, attempted to interpret or clarify the term "peaceful" so commonly used in the context of contemporary space activities.

While the world community, including the space powers, was urging that outer space be devoted exclusively to "peaceful" purposes, the United States and the Soviet Union were secretly developing and soon deploying satellites that were to serve a growing number of military objectives. A former chief scientist in the Pentagon reports that as early as 1955, the U.S. Air Force contracted the development of reconnaissance satellites; this and other U.S. military space programs at the time were proceeding "on a highly classified basis".[7] He says that U.S. "space programs, *from the beginning*, have been primarily of a military, not a civilian or scientific nature."[8] According to the same authoritative source, "both the U.S. and

---

[5] Quoted in M.S. McDougal, H.S. Lasswell and I.A. Vlasic, *Law and Public Order in Space* (New Haven: Yale University Press, 1963), p. 395.

[6] Ibid.

[7] Dr. Herbert F. York, "Nuclear Deterrence and the Military Uses of Space.", in *Weapons in Space*, F.A. Long, D. Hafner and J. Boutwell (eds). (New York: W.W. Norton. 1986), p. 17-19.

[8] Ibid., p. 20.

Soviet space programs overall have always been driven more by military considerations and requirements than civilian and scientific ones."[9]

It is not surprising, therefore, that as early as 1958 - 59, the legal position of the United States with respect to the meaning of the phrase "peaceful uses" became crystallized along lines quite dissimilar from the initial rhetoric. The term "peaceful" in relation to outer space activities was interpreted by the United States to mean "non-aggressive" rather than non-military.[10] Accordingly, all military uses are permitted and lawful as long as they remain "non-aggressive" as per Article 2 (4) of the UN Charter, which prohibits "the threat or use of force." By contrast, the Soviet Union publicly took the view, despite its own military uses of space, that "peaceful" meant "non-military" and that in consequence all military activities in outer space were "non-peaceful" and possibly illegal.[11]

II - Concept of "Peaceful Purposes" in Multilateral Treaties not Directly Related to Outer Space Activities

The use of the term "peaceful" is not limited to official outer space documents. In the post-World War II period, the term has appeared in several important multilateral agreements, namely, in the Statue of the International Atomic Energy Agency, the Antarctic Treaty, the Treaty for the Prohibition of Nuclear Weapons in Latin America, the Convention on the Prohibition of the Development, Production and Stockpiling of Bacteriological (Biological) and Toxin Weapons and on their Destruction and, most recently, in the UN Convention on the Law of the Sea.

The *Statute of the International Atomic Energy Agency*, adopted on October 26, 1956, stipulates in Article II that the objectives of the agency are to "seek to accelerate and enlarge the contribution of atomic energy to peace, health and prosperity throughout the world" and to "ensure ... that assistance provided by it or at its request or under its supervision or control, is not used in such a way to further any military purposes". Article II, enumerating the functions of the agency, also contains several references to the application of atomic energy for "peaceful" uses and purposes. There can be little doubt, especially when the negotiations preceding the adoption of the statute are taken into account, that the drafters of this agreement understood "peaceful uses" to mean non-military, rather than non-aggressive.

Of considerably more relevance to the context of outer space, chronologically the next multilateral agreement to employ the term "peaceful purposes" was the *Antarctic Treaty* of December 1, 1959. The first sentence of Article I declares that "Antarctic shall be used for

[9] Ibid.

[10] See, e.g., McDougal et al., Law and Public Order, No. 5, pp. 397 - 99.

[11] For more than twenty years scholars of international law in the Soviet Union have unanimously stated that 'use for peaceful purposes' should be interpreted as 'nonmilitary use'". P. Donay, "The Military Use of Outer Space: Implication for International Law", in H.G. Branch, (ed): *Military Technology, Armaments Dynamics and Disarmament* (London: Macmillan, 1989), pp. 471 - 75.

peaceful purposes only". The next sentence explicitly prohibits "any measures of a military nature ... as well as the testing of any type of weapons" on the continent. An official U.S. document credits the treaty with the "demilitarization" of the Antarctic and notes that it is based on "the premise that to exclude armaments is easier than to eliminate or control them once they have been introduced".[12] This treaty has often been invoked as the most authoritative aid for the interpretation of the term "peaceful" found in various outer space official texts.

The *Treaty for the Prohibition of Nuclear Weapons in Latin America*, signed on February 14, 1967, is another multilateral agreement in which the term "peaceful" signifies non-military. The contracting parties undertook "to use exclusively for peaceful purposes the nuclear material and facilities ... and to prohibit and prevent" on their national territories the "testing, use, manufacture, production or acquisition by any means whatsoever of any nuclear weapons" [Article 1 (1)]. Although Article I refers to nuclear weapons, the preamble of the treaty speaks of the "military denuclearization" of the region, a phrase that is broader in scope, especially when read in conjunction with Article 17, which provides for the right of the contracting parties "to use nuclear energy for peaceful purposes, in particular for their economic development and social progress".

The *Biological Weapons Convention*, signed on April 10, 1972, refers to "peaceful purposes" no less than six times in three different articles (I, II, and X). The term "peaceful purposes" in this Convention means that the use of toxins and biological agents is limited exclusively to prophylactic purposes and to the promotion of the "development and application of scientific discoveries in the field of bacteriology (biology) for the prevention of disease or for other peaceful purposes." [Article X (1)]. The development, production and stockpiling of toxins and biological agents "for hostile purposes" is prohibited; parties to the Convention undertake to destroy all their existing stocks of such agents (Article I and II).

The most recent appearance in a non-space multilateral agreement of the phrase "peaceful purposes" can be found in the *UN Convention on the Law of the Sea*, signed on October 7, 1982. Article 88, the shortest in this gigantic document, prescribes laconically that the "high seas shall be reserved for peaceful purposes". In this instance, the term "peaceful" most certainly does not mean "non-military," given the well-known fact that the high seas are navigated by naval vessels of many nations and used for tests of nuclear missiles as well as for naval maneuvers. Hence, it is difficult to find the rationale for the inclusion of the reference to "peaceful purposes" under the heading of "high seas"; it is no less difficult to explain the purpose of that part of Article 58 that declares that Article 88 also applies to the Exclusive Economic Zone. Similarly, it is not clear why Article 141 provides that the "Area" (i.e., the seabed and subsoil beneath the high seas) shall be "open to use *exclusively* for peaceful purposes by all States ..." However, here the use of the term "peaceful purposes" could be justified by reference to the widely accepted *Treaty on the Prohibition of the Emplacement of Nuclear Weapons and Other Weapons of Mass Destruction on the Seabed*

---

[12] U.S. Arms Control and Disarmament Agency, *Arms Control and Disarmament Agreements* (Washington, DC: 1982), p. 19.

and the Ocean Floor and in the Subsoil Thereof (of February 11, 1971) beyond a 12-mile coastal zone.

III - Concept of "Peaceful Purposes" in Multilateral Treaties Regulating Outer Space Activities

By the time negotiations on the *Outer Space Treaty* (OST) commenced, the United States and the Soviet Union were both irrevocably committed to using outer space for a variety of military purposes, especially for surveillance, communications, navigation and detection of nuclear explosions in space. The only significant difference in this respect between the superpowers was that the United States in contrast to the USSR made no secret of its military space programs. Although the Soviet Union continued publicly to treat military activities in outer space with disfavor, by accepting the text of the Treaty it implicitly acquiesced in the U.S. interpretation of the term "peaceful" in the context of outer space uses.

Contrary to the expectations of many states, the Treaty failed to ban all military uses of outer space. Under Article IV, paragraph one of the treaty, states shall not place "in orbit around the earth any objects carrying nuclear weapons or any other kind of weapons of mass destruction, install such weapons on celestial bodies, or station such weapons in outer space in any other manner". The treaty fails to give a definition of "weapons of mass destruction," a phrase that is commonly understood to mean, in addition to nuclear weapons, radiological, bacteriological, and chemical weapons, as well as any future weapons possessing large-scale destructive potential.

The second paragraph of Article IV stipulates that the "moon and other celestial bodies shall be used by all States Parties to the Treaty exclusively for peaceful purposes." The same paragraph explicitly bans the setting up of "military bases, installations and fortifications, the testing of any type of weapons and the conduct of military manoeuvres on celestial bodies." It will be noted that the "peaceful purposes" clause applies only to the moon and other celestial bodies but not to "outer space". According to former Legal Adviser in the U.S. Department of State, the "language of Article IV was carefully chosen to ensure that general principle of "peaceful uses" would not interfere with the testing" of weapons such as nuclear ballistic missiles.[13]

The adjective "peaceful" appears also in Article IX; any state party to the treaty "which has reason to believe that an activity or experiment planned by another State Party in outer space ... would cause potentially harmful interference with activities in the peaceful exploration and use of outer space ... may request consultation concerning the activity or experiment". Although no state has ever requested the consultations provided for by this article, conceivably it could be employed in restraint of certain potentially harmful military activities.

---

[13] A. Chayes, A.H. Chayes, and E. Spitzer, "Space Weapons: the Legal Context," in *Weapons in Space*, No. 7, pp. 193 -197.

The 1979 *Agreement Governing the Activities of States on the Moon and Other Celestial Bodies* repeats, in Article III, much of Article IV of the Outer Space Treaty. Article III of the Agreement prohibits the threat or use of force or any other hostile act on the moon and the use of the moon to commit such an act in relation to the earth or to manufactured space objects. This document adds little, if anything, to the provisions of the Outer Space Treaty relating to military space activities. Moreover, the fact that 10 years after its adoption it has received only a handful of ratifications, not a single one by any space-launching power, makes it at least for the time being irrelevant to the problem of controlling the military presence in outer space.

The most recent reference to "peaceful purposes" in a multilateral treaty can be found in the *Space Station Agreement,*[14] signed September 29, 1988 in Washington, DC. Article 1 of this Agreement States that its object is to establish a long-term international cooperative framework among the Partners ... for the detailed design, development, operation, and utilization of a permanently manned civil Space Station for peaceful purposes, in accordance with international law." As of November 1990, the Agreement is a long way from implementation.

The continuing degradation of the global environment, especially of the ozone layer and ionosphere, makes the *Convention on the Prohibition of Military or Any Other Hostile Use of Environmental Modification Techniques* (signed on May 18, 1977) of particular interest in the context of military uses of outer space. Although environmental modification techniques (ENMOD) are still very primitive and of scant interest to the military, the Convention's importance lies in its preventive effect; it restrains potential users of such techniques from initiating this type of activity. The Convention refers to "peaceful purposes" in the preamble and in Article III. In both instances it is obvious that "peaceful" is used in opposition to "military." Thus the preamble notes that the "use of environmental modification techniques for peaceful purposes could improve the interrelationship of man and nature," whereas "military or any other hostile use of such techniques could have effects extremely harmful to human welfare." The prohibited "techniques" are defined as "any techniques for changing -- through the deliberate manipulation of natural processes -- the dynamics, composition, or structure of the earth, including its biota, lithosphere, hydrosphere, and atmosphere, or of outer space" (Article II). An "understanding" relating to Article II, which is part of the negotiating record, includes among examples of ENMOD techniques "changes in the state of the ozone layer, and changes in the state of the ionosphere".

One noticeable weakness of this treaty is that it does not prohibit all major activities that could adversely affect the earth environment. Non-military ENMOD experiments could be as damaging as those intended for military or hostile purposes. The Convention explicitly stipulates that its provisions should not "hinder the uses of environmental modification

---

[14] "Agreement Among the Government of the United States of America, Governments of Member States of the European Space Agency, the Government of Japan, and the Government of Canada on Cooperation in the Detailed Design, Development, Operation and Utilization of the Permanently Manned Civil Space Station." Full text in *United States Space Law - National and International Regulation,* Release 89-1, Jan. 1989, p. 3.

techniques for peaceful purposes," which means that experimentation aimed at the altering of natural phenomena may continue. Yet any major artificially induced disruption of weather patterns is inherently hazardous, with potentially catastrophic consequences and of still unproven socioeconomic utility. The permissive provisions of the Convention could also surreptitiously be used for prohibited, i.e., military, purposes either in the guise of "peaceful utilization" of ENMOD current techniques, or for supposedly prophilactic and defense purposes (just as continuing research and development of biological agents is being justified). Of course, this points to the more general difficulty in drawing an effective line between permitted and prohibited research that might relate to military uses. The problem is especially acute in the ENMOD setting, however, where permissible experimentation is not limited to the laboratory. A second concern is that recourse to the Consultative Committee of Experts, provided for in Article V to assist in the solution of problems arising out of the application of the Convention, is not mandatory. The effectiveness of the Convention would have been greatly enhanced had the text provided for the mandatory screening by experts of experiments while still in the planning stage.

Although the term "peaceful" does not appear in the *Limited Test Ban Treaty* of August 5, 1963, this agreement must be mentioned not only because it expressly extends to outer space the ban on nuclear explosions of *any* kind, but also because of its exceptional place in the modest corpus of arms limitation agreements of universal application. Under Article I, each state party undertakes "not to carry out any nuclear weapon test explosion, or any other nuclear explosion, at any place under its jurisdiction or control: (a) in the atmosphere; beyond its limits, including outer space; or under water, including territorial waters or high seas ..." Even in terms of the number of states parties, this treaty is the most successful disarmament agreement on record, with well over 100 ratifications.

IV - The Meaning of "Peaceful Uses" in Current International Space Law -- Conclusions

Following the adoption of the Outer Space Treaty, efforts to provide a "correct" interpretation of the term "peaceful" as employed in the Treaty intensified. This was prompted primarily by the continuing and rapid expansion of military space activities by the two superpowers that were also the main architects of the treaty. Whereas the results of the many exercises in interpretation remain inconclusive to this day, through their conduct before and especially after the conclusion of the treaty, the two leading users of outer space provided the legal meaning to the term "peaceful." Under the 1969 *Vienna Convention on the Law of Treaties*, the most authoritative text concerning the interpretation of international agreements, words in a treaty must be interpreted in accordance with their "ordinary meaning" [Article 31(1)]. In all reputable dictionaries of the English language, the term "peaceful" is defined much as in the Oxford English Dictionary where "peaceful" means "disposed or inclined to peace; aiming at or making for peace; friendly, amicable, pacific".[15] By no stretch of the imagination can this description be applied to any current or past military use of space. If

[15] *The Compact Edition of the Oxford English Dictionary*, II (New York: Oxford University Press 1971), p. 2105.

"peaceful" means "non-aggressive," then it follows logically -- and absurdly -- that all nuclear and chemical weapons are also "peaceful," as long as they are not used for aggressive purposes.

However, Article 31, paragraph 3 of the Vienna Convention provides that in the process of interpretation "any subsequent practice in the application of the treaty" shall also be taken into account. Discussing conditions that must be fulfilled before a conventional rule can be considered to have become a rule of customary law, the International Court of justice declared in the landmark *North Sea Continental Shelf Cases* that in addition to widespread practice and representative participation in a convention, the practice must include states whose interests are "specially affected."[16] In our case, with only the Soviet Union and the United States active in outer space before and for some time after the entry into force of the OST, the "practice" of even one space power, clearly a "specially affected" state, carried substantial weight in law. All the more so when supported by several other states with developing space capabilities.

Given the ambiguity of the term "peaceful" as used in the OST, as well as the overt and covert practice of the two state actors in outer space, the conclusion is inescapable that all military uses of space other than those prohibited by treaty were -- since the beginning of space exploration and are still today -- lawful as long as they do not violate any of the principles and rules of general international law (e.g., uses that represent the threat or employment of force). Furthermore, it is a well-established rule of international law that in order to prevent a particular interpretation of a conventional rule from becoming controlling, dissatisfied states must signify their disagreement formally, either through diplomatic channels or through public statements of authoritative government spokespeople (as has, e.g., the United States with respect to Libya's claim of sovereignty over the Gulf of Sidra, or the United States and the USSR with respect to claims by seven states to sovereignty over parts of Antarctica). No state has ever *formally* protested the U.S. interpretation of the phrase "peaceful uses" in the context of outer space activities.

In practice, the military presence of the two superpowers in space grew so rapidly that not long after the adoption of the OST, outer space achieved the dubious distinction of being the most heavily militarized environment accessible to humans (based on the number of military and civilian payloads launched into orbit).[17] According to the Commander-in-Chief of the U.S. Space Command, between 1957 and 1988 the USSR conducted "nearly 2,000

---

[16] [1969] I.C.J. Reports 3, para. 73. A rule becomes a rule of customary international law when a significant majority of states, including states whose interests are specially affected, act in accordance with that rule because they believe it to be binding. In the application of such a rule state practice must be both extensive and virtually uniform.

[17] For an overview, see P.B. Stares, *The Militarization of Space: U.S. Policy, 1945-1984* (Ithaca, NY: Cornell University Press, 1985); B. Jasani, (ed), *Outer Space -- A New Dimension of the Arms Race* (London, Taylor and Francis for SIPRI), 1982); P.B. Stares, "U.S. and Soviet Military Space Programs: A Comparative Assessment," in *Weapons in Space*, No. 7, p. 127.
"The militarization of space is an accomplished fact -- on both sides ... Now a new phase is beginning -- the *weaponization* of space." R.M. Bowman, *Star Wars* (Los Angeles: Jeremy P. Tarcher, 1986), pp. 8 - 9.

space launches, at least 90 % of which have had military-related missions."[18] During the same period, at least 50 percent of all U.S. launchings have served military purposes. Even if the claim about the Soviet effort is an exaggeration, there can be no doubt that the military presence in space has been and remains the dominant feature of this environment.

It has been suggested that the term "peaceful" in the Outer Space Treaty should be reserved only for non-offensive space instrumentalities, i.e., for civilian spacecraft and military space hardware other than space weapons.[19] Accordingly, anti-satellite devices and various weapons in the process of research and development as part of ballistic missile defenses would not qualify as "peaceful." As a *de lege ferenda* proposal, this suggestion merits attention; as a statement purporting to express the current position of space law, it is inaccurate. At this time, there is no clearly enunciated principle or rule in international law to prevent any state, except the United States and the USSR bound by the ABM Treaty, from deploying (not just developing and testing) a whole array of weapons such as those conceived in the U.S. Strategic Defense Initiative program. Only the space-based X-ray laser, designed to channel the energy of an exploding nuclear weapon into beams of radiation, would fall under Article IV of the Outer Space Treaty in the category of prohibited uses of outer space; this prohibition is, most likely, already part of customary international law given the large number of states, including all the nuclear weapon states, that are bound by the OST.

In the same limited category of prohibited uses, one should include biological, though not necessarily chemical, weapons. The possession of chemical weapons is not unlawful in international law; there is no chemical weapons equivalent of the widely accepted *Biological Weapons Convention.* Only when a treaty banning the production and possession of chemical weapons is concluded and ratified by a large number of states, particularly by major military powers, it should be possible to assert that the stationing of such weapons in outer space would be unlawful both in treaty and customary law.

However, this is not to say that the legality of some weapon systems envisaged in the U.S. Strategic Defense Initiative could not be questioned, as illustrated in the following section. Nevertheless, because only a widely endorsed multilateral treaty or a declaration by the World Court could authoritatively designate a particular weapons system or activity in outer space to be unlawful, given the unlikelihood of any such treaty and the traditional reluctances of states to resolve their differences through international adjudication, the legal characterization of such uses could remain uncertain for a long time.

Perhaps the most important lesson that can be drawn from the above survey and one that should be strongly impressed on governments is to avoid the imprecise term "peaceful"

---

[18] J. Piotrowsky, "Space: The Unseen Force", *Defense* (Feb. 1988), pp. 132 - 135. According to General R.T. Herres, former commander of the U.S. Space Command, in 1986 about 95 percent of Soviet launches had military applications. R.G. O'Lone, "USAF Official Calls Soviet Satellites Threat to U.S. Carrier Battle Groups", *Aviation Week and Space Technology*, Sept. 29, 1986, p. 20. Only in 1985 did the Soviets publicly acknowledge having a military space program.

[19] E.F. Hennessey, "Liability for Damage Caused by the Accidental Operation of a Strategic Defense Initiative System", *Cornell International L.J.*, 21 (1988), pp. 317-328.

in all future arms limitation and disarmament agreements, unless the term is defined in each treaty with great precision. Up to now, the term, ironically, has served merely to engender controversy and confusion while adding nothing to the noble goals of peace and disarmament.[20]

V - Military Uses of Outer Space that Constitute a Threat to International Peace and Security

Rather than rely on the unworkable distinction between "peaceful" and non-peaceful uses of outer space, future international agreements designed to limit or prohibit certain uses of, or weapons designed for employment in, outer space should explicitly name the weapon or activity to be banned or restricted. What follows is an attempt to catalogue all military space uses, current as well as those known to be in the planning phase, under three headings: (1) uses that are prohibited by treaty or customary law, (2) uses that are lawful under current space law, (3) uses that should be banned because of their destabilising effect on international peace and security. In this section, uses belonging to categories 1 and 3 are identified.

Uses (Activities) Prohibited by Treaty or Customary International Law

1. Placing nuclear weapons in orbit around the earth or on celestial bodies or anywhere else in outer space (Article IV, OST; Article III, *Moon Treaty*). This would include nuclear-armed satellites functioning as space "mines."
2. Placing weapons of mass destruction in orbit around the earth, on celestial bodies or anywhere else in outer space. It is generally understood that the reference to "weapons of mass destruction" includes biological agents, now expressly prohibited by the Biological Weapons Convention, radiological weapons, as well as any future weapons whose destructive potential would be "catastrophic" (Article IV, *OST*).
3. The establishment of military bases and installations, the testing of any kind of weapons, and the conduct of military maneuvers on the moon and other celestial bodies (Article IV, paragraph 2, OST; Moon Treaty, Article III).
4. Carrying out any nuclear weapon explosions, or any other nuclear explosion, anywhere in outer space (Limited Test Ban Treaty, Article I.1.(a); customary law).

[20] See, e.g., contradictory "Comments" in the *International Space Law Committee of the International Law Association's Proceedings of the Warsaw Conference (1988)* on "The Continuing Conflicts in the Interpretation of the Legal Rules Governing Military Space Activities". One of the commentators, Professor S. Gorove, put it this way: "Because the term 'peaceful' has been subject to two diametrically opposed interpretation, one meaning 'nonmilitary' and the other 'nonaggressive,' there is no assurance for the uniform application of the law whenever reference is made to 'peaceful' uses in the space treaties". To avoid future controversy, the author suggests that the drafters of any new arms limitation or disarmament agreement should agree on the "more precise identification of prohibited or permissible activities in lieu of a general reference to 'peaceful' uses." Ibid., p. 27.

5. Military or hostile uses of environmental modification techniques that could produce widespread adverse effect on the human environment, which includes both the earth's atmosphere and the surrounding outer space (ENMOD Convention, Articles I and II).

6. Any hostile act, committed by a device designed to operate in outer space, that causes damage to the assets of another state located in outer space, or to foreign aircraft in flight or on the territory, including the airspace, of another state, or in the area outside the jurisdiction of any state (e.g., the high seas) (General international law; United Nations Charter, Article 2(4); UNGA Resolution 3314 (XXIV) of December 14, 1974, on the Definition of Aggression, Articles 3 and 4).[21]

7. Any intentional physical interference, whether or not resulting in damage, with space assets of another country located in outer space without that country's authorization (e.g., unauthorized inspection of another state's satellite) (General international law; OST Articles III, VI, VIII and IX).

Uses that should be Banned because of their Destabilizing Effect on International Peace and Security

1. The development and deployment of (1) co-orbital and (2) direct-ascent anti-satellite interceptors, or ASAT weapons of any kind. There is wide support in the world community for a ban on such weapons. Their capability to "impair or destroy satellites used for nuclear command and control could seriously undermine strategic stability in periods of heightened tension between the superpowers",[22] notes an informed observer. An international agreement prohibiting ASAT weapons should provide also for the destruction of existing ASAT weapons, subject, of course, to appropriate verification.

2. The testing in outer space of "dedicated" as well as any other kind of ASAT devices. A recent comprehensive study on the state of the space environment begins with the assertion that the "greatest hazard facing human activities in outer space today is space debris".[23] And according to a Soviet official publication, there are some 40,000 objects the

---

[21] Although under the UN Charter resolutions of the General Assembly, with few exceptions, are merely recommendations, through the practice of member states certain resolutions, dealing with important legal issues, which have been adopted with overwhelming majorities, including all the major powers, have come to be regarded by states as amounting to much more than a mere recommendation. Whereas their legal status in international law is difficult to define, it suffices to note that states often treat such resolutions as if they contained legal obligations. There are perhaps not more than a dozen UNGA resolutions pertaining to this very special category of legal texts, and the resolution defining aggression is one of them. The 1963 Declaration of Legal Principles governing the Activities of States in Outer Space and the 1948 Universal Declaration of Human Rights are additional examples of such trail-blazing UNGA resolutions. Their role is particularly important in the process of ascertaining the existence of a customary rule of international law.

[22] B.K. Maclaury, Foreword, in P.B. Stares, *Space and National Security*, (Washington, DC: The Brookings Institution, 1987), p. ix.

[23] H.A. Baker, "Space Debris: Law and Policy in the United States", *University Colorado Law Rev.*, 60 (1989) 55; see also E. Vitt, "Space Debris", *Space Policy*, 129 (May 1989), 129; H.A. Baker, *Space Debris: Legal and Policy Implications* (Boston: Martinus Nijhoff, 1989).

size of a tennis ball or larger in near-earth space.[24] ASAT tests could greatly add to the volume of debris in the most heavily used part of space, causing interference with both lawful civilian and military uses. It is not inconceivable, furthermore, that accidental collision between a satellite and debris might be mistakenly interpreted as the deliberate use of a space weapon.[25]

3. All weapons, regardless of basing, specifically designed to attack targets in outer space. Such a ban would protect both civilian satellites and unarmed military space assets.

4. It has been reported that phase one of the U.S. ballistic missile defense (BMD) might involve a deployment of "several thousand interceptors."[26] Although Article II of the OST declares the principle of freedom of outer space, the launching into space of an armada of space "battle stations" to assure that a large number of them is always over the territory of a target state would seem to violate the more important norm prohibiting the "threat of force" in relations between states [UN Charter, Article 2(4)]. In spite of claims that such a space-borne BMD system is purely for defensive purposes -- a shield against hostile nuclear-tipped missiles -- it is generally understood that this system could also serve the purpose of a nuclear surprise attack. Its deployment would be exceedingly destabilizing.[27] It is hard to believe that such a space-based BMD system would not provoke a diplomatic *and* military response, initially most likely by anti-satellite devices.

5. Satellites designed to cause electronic interference with the space assets of another country, thereby rendering them useless or materially impairing their value.

Military Uses of Outer Space in Accord with Contemporary International Law

Modern military establishments, primarily those of the leading superpowers, have come to rely heavily on outer space technology. Particularly crucial is the role of surveillance and communications satellites, not merely for the armed forces of these countries but also for monitoring compliance with arms limitation agreements as so-called national technical means of verification. An official U.S. survey stresses that "without satellites, performance of many military missions would become impossible, and performance of others would require large

---

[24] *Soviet Cosmonautics: Questions and Answers*, Valentin Glushko (ed), (1988), p. 39.

[25] B. Jasani and M. Rees, "The Junkyard in Orbit", *Bulletin of the Atomic Scientists*, 45 (October 1989), pp. 24 - 25.

[26] T.H. Johnson and A.B. Carter, "ASAT and BMD Weapons", in *Defending Deterrence: Managing the ABM Treaty Regime into the 21st Century*, Antonia H. Chayes and Paul Doty (eds), (Washington DC: Pergamon/Brassey's, 1989), p. 42. See also P. Mann, "Early Deployment Crippled by Lack of Advanced Launcher", stating that the SDI program will require "launching as much as 5 million lb. into orbit annually", including hundreds of satellites and thousands of small space- and ground-based missiles. *Aviation Week and Space Technology*, July 11, 1988, p. 38. The latest version of the space-based segment of the SDI envisages the launching into orbit of some 6,000 so-called Brilliant Pebbles, each 3 feet long and weighing 100 pounds. Time [Magazine], June 26, 1989, p. 26.

[27] See, e.g., G. Rathjens and J. Ruina, "BMD and Strategic Instability", in *Weapons in Space*, no. 7, p. 239.

increase in the unit strengths of various U.S. force elements".[28] The great potential of satellites for a variety of military functions was discovered soon after the advents of space age and that early discovery explains why military uses of space quickly acquired the status of a lawful activity.[29]

1. The use of satellites for communications, navigation, photoreconnaissance, gathering signals intelligence, ocean surveillance to locate and track warships, detection of nuclear explosions in the space and earth environments, ballistic missile early warning, and for weather monitoring is lawful under current international space law. The use of military personnel in outer space is also lawful (explicitly permitted by Article IV, paragraph 2 of the OST, in relation to the moon and other celestial bodies).

2. Space stations, even if serving exclusively military purposes, may be freely erected in outer space, with only minimal restrictions provided for in Article IV of the OST.

3. Under the present legal regime, future aerospace planes, capable of speeds up to Mach 25 and already in the advanced research stage in several countries (United States, USSR, Japan, F.R.G., United Kingdom, and France), will be able to operate in space subject only to modest legal restrictions. Indeed, the main legal problem facing this instrumentality would be how to circumvent the well-established principle of air sovereignty.

4. Although hardly fitting the adjective "peaceful," anti-satellite weapons of every kind, except those nuclear-armed and space-based, are not unlawful under current international law.

5. In the absence of an explicit treaty prohibition, states are free to use measures such as camouflage and deception to conceal sensitive military activities in outer space (so-called passive measures).

6. For quite some time nuclear substances have been used in spacecraft to provide the energy for various instruments aboard. This form of energy is regarded as much superior to other sources of energy for space missions requiring huge amounts of power and deep-space missions. Directed energy weapons would rely heavily on nuclear power because of their need for large amounts of energy. Of course, nuclear-powered satellites occasionally malfunction and crash-land on earth. The impact of *Cosmos 954* on the territory of Canada, in January 1978, has led to demands for strict international regulation of nuclear power sources in space, even for their total ban.[30] As long as the number of nuclear-powered satellites is small, no question of their legality is likely to arise, provided, of course, that they

---

[28] US Congress, Office of Technology Assessment, *Anti-Satellite Weapons, Countermeasures, and Arms Control* (Washington, DC: US Government Printing Office, 1985), 34. See also *National Security Strategy of the United States*, where space systems are described as a "vital element of US national power". Department of State Bulletin, April 1988, 1, p. 18.

[29] This position with respect to the legality of military uses of outer space was not shared by the Eastern bloc publicists. See Donay, "Military Uses". See esp., M. Lachs, *The Law of Outer Space: an Experience in Contemporary Law-making*, (Leiden: Sijthoff, 1972), pp. 106 - 8; M. Marcoff, *Traité de droit international public de l'espace*, (Fribourg: Editions Universitaires Fribourg Suisse, 1973), p. 357.

[30] Soon after the Cosmos 954 incident, the UN Committee on the Peaceful Uses of Outer Space began drafting a set of principles designed to regulate the use of nuclear power sources in outer space. The purpose is to assure the safe use of nuclear energy on board space objects. For an up-to-date report, see UN Doc. A/AC.105/430, 26 April, 1989 (Report of the Legal Subcommittee of COPUOS).

cause in the meantime no catastrophic accidents. However, a large number of such satellites in near-earth orbit, especially if they serve military functions, could lead to demands for severe restrictions on their deployment and even for their prohibition because of the perceived threat they represent.[31]

7. The legality of launching into space, especially in the more congested areas, a large number of decoys could also be questioned at some point in the future. Doubts concerning legality of this practice could arise either because such decoys cause interference with the lawful activities of other users of outer space, or because they make difficult the implementation of an arms limitation agreement by engendering confusion. At present, there are no restrictions in law on the number of objects a state may launch into outer space.[32]

VI - Monitoring Outer Space Uses -- Verification

The *Final Document* of the First Special Session of the UN General Assembly on Disarmament, held in 1978, declared that "to prevent an arms race in outer space, further measures should be taken and appropriate international negotiations held in accordance with the spirit of the ... [Outer Space] Treaty".[33] Seven years elapsed before the question of military uses of outer space was placed in 1985 on the agenda of the Conference on Disarmament (CD). Regrettably, not much progress toward the goal of making outer space free of arms has been registered to date in this multilateral forum. In the parallel bilateral negotiations between the United States and the Soviet Union, designated as "Defense and Space Talks," similarly little progress has been achieved on space subjects.[34] Whereas these negotiations are conducted separately, they are nevertheless closely interrelated in terms of

[31] The eminent Soviet scientist, S. Rodionov, considers all space-based nuclear energy sources as a threat both to the terrestrial environment and to other satellites in orbit. He writes: "Space-based nuclear reactors are a good example of activity which has a very negative effect on international security and stability. Such activity should be banned". "State of Research on Disarmament and International Security questions related to Outer Space", *UNIDIR Newsletter*, No. 3, September, 1989, pp. 7 - 9.

[32] The view that "everything not directly prohibited in the sphere of military activity in outer space is lawful contradicts the general principles of international law, trends in the legal regulation of activities of this type, the world's sense of legality and international moral norms," writes a leading Soviet commentator, Dr. V. Vereshchetim in *Prevention of the Arms Race in Outer Space* (New York: UNIDIR, United Nations Publications, 1986) p. 111. Even if one agrees with this view, and many would, one cannot ignore the practice of states, especially in outer space, which has been in the other direction.

[33] United Nations, Department of Public Information, "Final Document -- Special Session of the General Assembly on Disarmament," (23 May -- 1 July 1978), para. 80.

[34] In these talks, according to President Bush, the U.S. objective "will be to preserve our options to deploy advanced defenses when they are ready". Address at Texas A & M University, 12 May, 1989. U.S. Department of State, Bureau of Public Affairs, *Current Policy* No. 1175, p. 2. For an up-to-date report on these negotiations, see H.F. Cooper, "The Defense and Space Talks Small Steps towards Agreements", *NATO Review*, No. 4 (August 1989), p. 11. In a speech delivered on 24 May, 1989, the president again stressed the U.S. commitment "to deploy when ready -- a more comprehensive defensive system, known as SDI." U.S. Department of State, Bureau of Public Affairs, *Current Policy* No. 1178, p. 2.

subject matter. Success in one should have a positive impact on the other. In either case the end result should appear in a treaty that would incorporate agreed restrictions on the military uses of outer space, e.g., a ban on ASATs. Should the U.S.-USSR bilateral negotiations result in a treaty banning a certain weapon system, the treaty should be open for accession to other states, as are the Limited Test Ban Treaty and the Non-Proliferation Treaty. There can be no doubt that a large majority of countries would become party to such a treaty leading eventually to the emergence of a customary rule binding upon all states.

The contemporary trends in the militarization of outer space have prompted the UN General Assembly to conclude that the legal regime applicable to outer space, as such, is not sufficient to guarantee the prevention of an arms race in outer space (Resolution 42/33 of December 1987). The Assembly issued an urgent call for new measures, " with appropriate and effective provisions for verification", to consolidate that regime and to enhance its effectiveness. In earlier sections of this essay, the uses of space that are already prohibited by international law, as well as those uses that in the opinion of this author should be banned, were identified. What remains to be dealt with, however briefly, is the verification aspect of possible future agreement restricting or prohibiting specific weapons systems or particular uses of outer space.

There are two key issues in verification -- the technology and the proper scope of verification. Although the current state of verification technology is quite advanced, consisting of, *inter alia*, photographic, infrared, radio frequency, and microwave detectors, even more advanced devices will be needed to monitor space-based weaponry. In this connection, it is worth noting that Canada's PAXSAT proposal claims that "the design of all spacecraft and ... their orbital parameters, together with the nature of signals to and from the spacecraft, can provide highly significant data as to that spacecraft's function".[35] Using only a space-based observation system, according to the PAXSAT feasibility study, such a system should be able to monitor the function of a spacecraft by co-orbiting and keeping stationary with the target satellite or even by a fly-by.[36] It is also noteworthy that during the inconclusive U.S.-USSR ASAT talks in 1978 - 1979, the Soviet negotiators claimed that a ban on such weapons would be verifiable by national technical means.[37] Still, as a recent report released in the United States observes, in the near future it will be necessary to monitor continuously the conduct of suspicious objects in space, because "space weapons, whether space-based or ground-based, may give little evidence of their true nature once on station".[38]

As to the scope of verification measures, by far the most effective and most reliable would be on-site, pre-launch inspection of all objects, both civilian as well as military, destined for outer space. This is the kind of verification not provided for in the Registration Convention of 1975, an agreement that is so permissively drafted as to be virtually useless

---

[35] External Affairs, Canada, *Paxsat Concept* (1986), p. 41.

[36] Ibid.

[37] UNIDIR, *Disarmament: Problems related to Outer Space* (New York: United Nations, 1987), p. 181.

[38] T.M. Foley, "Monitoring Soviet Space Weapons Adds to Demand for U.S. Intelligence", *Aviation Week and Space Technology* 27 February, 1989, p. 22.

for the purposes of any meaningful monitoring of objects launched into outer space.[39] Current proposals advocating an amendment to the Convention that would require states parties to provide more accurate and more timely information about the function and characteristics of each satellite would only modestly increase our knowledge about space activities but would have no real effect on the arms race in space. Indeed, negotiating such a minor agreement could significantly delay the adoption of genuine arms reduction and disarmament measures.

Yet an all-embracing, comprehensive verification procedure may not be entirely beyond reach. The unprecedented scope of verification provisions incorporated in the U.S.-USSR *Treaty on the Elimination of their Intermediate Range and Shorter Range Missiles* (INF Treaty), notably the highly intrusive character of the inspection system agreed upon, augurs well for future agreements on outer space matters.[40] Furthermore, in the on-going negotiations on the convention banning chemical weapons, agreement has already been reached in principle on an inspection system even more intrusive than that of the INF Treaty.[41]

Dramatic changes in the Soviet approach to verification can be seen when comparing the monitoring clauses of the 1981 Soviet proposal of a treaty banning the stationing of weapons in outer space, with the INF Treaty and the more recent Soviet statements relating to space arms. In the 1981 draft proposal, verification of compliance was to be limited to each party's "national technical means".[42] In March of 1987, the Soviet Union proposed in the CD a drastically different international verification system within the framework of a treaty banning space weapons. Under this regime, inspectors would have the right of access "for the purpose of on-site inspection, to all objects destined to be launched and stationed in space, and to their corresponding launch vehicles".[43] Later in the same year, the Foreign Minister of the Soviet Union expanded on the original proposal by stressing that the USSR advocates "a permanent presence of groups of inspectors at all space launch sites".[44] In addition, the Soviet Union is ready to provide the inspectors with advance notice of every launch, with information about the object to be launched; should there be an agreement

[39] For some interesting suggestions concerning the adaptation of the Registration Convention to the needs of an ASAT verification regime, see UNIDIR, *Satellite Warfare: A Challenge for the International Community* (Report by the French Institute for International Relations, United Nations, New York, 1987), p. 27.

[40] The INF Treaty contains such sweeping measures as short-notice, on-site inspections, continuous monitoring of missile production facilities, and detailed exchange of information. See INF Treaty, articles VIII, XIII.

[41] See, e.g., J. Goldblat, "Chemical Disarmament: From the Ban on Use to a Ban on Possession", (Canadian Institute for International Peace and Security, *Background Paper* No. 17, February 1988); T. Bernauer, *The Projected Chemical Weapons Convention: A Guide to the Negotiations in the Conference on Diarmament* (UNIDIR, United Nations, New York, 1990).

[42] Article 4, "Draft Treaty on the Prohibition of the Stationing of Weapons of any kind in Outer Space". UN Doc. A/36/192, Annex (20 August 1981).

[43] *Disarmament: Problems related to Outer Space*, n. 31, p. 164.

[44] Ibid.

reached on a total ban on space weapons, inspections would extend to laboratories and industrial plants.[45]

The new openness practiced by the Soviet Union is in some measure being reciprocated by the United States. In a foreign policy speech delivered May 12, 1989, President George Bush recalled President Eisenhower's "Open Skies" arms control proposal of 1955, which provided for the aerial inspection of the Soviet Union by U.S. aircraft and of the United States by Soviet aircraft.[46] President Bush appealed to the Soviet Union to explore urgently with the United States the "Open Skies" proposal, "but on a broader, more intrusive and radical basis -- one which ... would include allies on both sides". He invited all interested countries to "meet soon to work out the necessary operational details ... Such surveillance flights, complementing satellites, would provide regular scrutiny for both sides. Such unprecedented territorial access would show the world the meaning of the concept of openness".[47] By September 1989, the Soviet government informed the United States of its willingness to give "active support" to Bush's proposal and by February 1990, negotiations were in progress with 16 member States of NATO and 7 Warsaw Pact Nations participating.[48]

VII - Conclusion

As the above survey suggests, reaching agreement on military uses of outer space will not be easy or simple. The "bewildering complexity of the problem" involved both in preventing the deployment of space weapons and in "defining the kinds of military activities that might

[45] Ibid., p. 165. For a comprehensive survey of Soviet policies in regard to verification, see M. Kokeyev and A. Androsov, *Verification: the Soviet Stance - Its Past, Present and Future* (UNIDIR, United Nations, New York, 1990). The latest and up to now perhaps the most remarkable illustration of Soviet "glasnost" as applied to military secrets has been the opening to U.S. visitors of Sary Shagan testing ground for laser research. According to the Pentagon, this facility is the "core of the Soviet anti-missile 'Star Wars' program". B. Keller, "American Team Gets Close Look at Soviet Secret", *New York Times*, 9 July, 1989, p. 1, c.5.

[46] *Current Policy*, n. 28, p. 2.

[47] Ibid.

[48] M. Gordon, "Shevardnadze Gives Bush Arms Offer as Summit Plans are Worked out", *New York Times*, 22 September, 1989, p. A10, c.1. As of December 1989, some important details of the proposed "Open Skies" system are not known. Thus, e.g., it is not clear whether the system will cover the territory of all member states of the two alliances; whether the information gathered will be widely or restrictively disseminated; and which countries will be allowed to carry out aerial inspection. What is known is that the scheme would not be related to any specific arms control agreement but would rather serve as a confidence-building measure. See J. Boulden, "1990 Open Skies Conference in Canada", *Peace & Security* (Canadian Institute for International Peace and Security), 4 (Winter 1989/1990), 13.

not be legitimately conducted in space"[49] will require from the negotiating parties patience, good faith and, above all, respect for the genuine concerns of the community of nations.

One can only hope that the new atmosphere in East-West relations, which has already, through the INF Treaty, resulted in the elimination of an entire modern and expensive weapon system, will continue to improve and thus make it possible to satisfy the universal yearning for an outer space free of weapons and with a reduced military presence.

[49] "Ambassador Marchand Addresses CD on Prevention of Arms Race in Outer Space". A speech delivered on July 26, 1988. [Department of External Affairs of Canada], *The Disarmament Bulletin*, 8 (Summer 1988), 14.

CHAPTER 4: TECHNICAL ASPECTS OF PEACEFUL AND NON-PEACEFUL USES OF SPACE

Isabelle Sourbès and Yves Boyer

Space-based military systems may serve to stabilize confrontational situations in allowing better transparency between opponents. They also facilitate progress in arms control. Among civil applications, such as remote-sensing and communications, space may become an arena where various types of experiments could be carried out, the applications of which on earth would allow states to acquire an industrial or economic advance that would give them an overwhelming supremacy.

Thus whereas certain criteria allow differentiations between civil and military functions, attention should also be paid to force postures and doctrinal concepts governing the use of forces by states involved in the use of outer space.

Any characterization of what represents the technical aspects of peaceful and non-peaceful uses of space is an issue that lends itself to a varied treatment. There are indeed at least two ways to deal with it. One is to take a purely technical approach and describe the various types of satellites in assessing their performance and role. Limited in scope, this way of handling the topic remains very static and offers no other perspectives than examining and comparing facts, capabilities, and data. Another way to deal with this subject is much more ambitious. It devotes more attention to force postures and doctrinal concepts governing the use of forces and connects those concepts to activities developed in space. Accordingly, it will be possible to characterize those activities in relation to specific strategic situations. However, even in following this approach, one runs the risk of stumbling over a problem of definition because "peaceful or non-peaceful activities" are very subjective notions. A reconnaissance satellite may, for example, help to monitor the compliance with a treaty; it may also be used to prepare air strikes against specific targets.

The Outer Space Treaty (OST) signed by the United States and the USSR in 1967[1] used the notion of peaceful purposes for space activities. But already at that time, one could have witnessed that outer space was an area where most activities were driven by military rather than by scientific needs. At the time of the treaty signature, satellites had already been orbiting the earth for 10 years. Even before, a study written in 1946 under the aegis of what became known as the Rand Corporation forecasted the potential use of satellites for military applications. Some years later, in 1955, General James Gavin[2] made predictions about the

[1] In 1990 more than 100 countries had signed the Outer Space Treaty.

[2] As quoted by G.M. Steinberg, in *The Exploitation of Space*, M. Schwarz and P. Stares (eds) (London: Buttenworth, 1985).

decisive importance of reconnaissance satellites to gather data particularly for developing, in relation to nuclear deterrence, a comprehensive set of strategic targets located in the territory of potential adversaries. Accordingly, the first military satellites launched by the United States were for reconnaissance purposes. Initially, the Soviet leadership considered U.S. reconnaissance satellites as mere spying means and contemplated the possibility of destroying them or at least degrading their function. Despite this initial attitude, the USSR gradually recognized the importance of gathering information collected by satellite for strategic purposes and, in 1963, it followed the American initiative.

The OST a few years later could not consequently ban military activities that already existed for the greatest benefit of the superpowers. This discrepancy between the spirit of the treaty and the strategic realities explains in great part the choice of an ambiguous wording in the treaty such as "peaceful and non-peaceful activities." This ambiguity regarding whether the notion of militarization of space is peaceful or not has poisoned the debate since that time. The treaty in fact provided a set of general rules governing the use of space and could not be more specific in order to obtain a general consensus among nations at a time when, the space conquest just started, it was impossible to explore all its consequences in terms of international relations.

I - Overview

Issues raised by the interpretation of the meaning of peaceful or non-peaceful purposes of space activities are complex to clear up. Since the signature of the OST, interpretations of its clauses are still debated. Some analysts maintain that all military space systems must be banned. This was the approach first taken by the Soviets. Others argue that only "aggressive" space systems are the object of the OST. This has been, since the inception of the treaty, the interpretation followed by the United States (see Chapter 3).

This divergence continues today. The replacement of "peaceful" by "non-aggressive" does not bring clarification. The fact that growingly military satellites are integral parts of a force structure and a force posture as well as the dual capability of spacecraft whether they provide civilian or military uses complicates the problem.

The treaty raises other uncertainties, for example, the lack of definition of the boundary between the air space and the outer space (discussed in Chapter 2), as well as the lack of precision about the repartition of the common heritage of space resources. These missing definitions are due to the way the space law has been elaborated under the UN-aegis and to the constant technological progress of space activities, which go faster than the process of regulation.

When using the treaty to evaluate the peaceful feature of space activities, it appears that the treaty mentions global principles, which are peaceful purposes, and provides the substance of the arms control provision in only one article. Its first alinea prohibits only the launching of mass destructive weapons in space and does not take into consideration military satellites, whereas the second alinea insists on the idea of the "use of the moon *exclusively for peaceful purposes*." But, even if that case is more restrictive concerning military activities, military resources are not excluded. Article IV says: "use of military personnel for scientific research or for other any peaceful purposes shall not be prohibited. The use of any equipment or facility necessary for peaceful exploration of the moon and other celestial bodies shall also not be prohibited."

In fact, the developments of both military and civilian programs have always been related to each other as demonstrated in history. Since the end of World War II and almost 10 years before the recommendation made by the organizing Committee of the International Geophysical Year, the Soviet Union as well as the United States have been engaged in the development of intercontinental ballistic missiles. This non-peaceful approach led to the first launch of a Soviet satellite Sputnik by the intercontinental rocket R-7.

In the context of the Cold War, the launch of Sputnik opened the race to the militarization of space to the detriment of purely scientific activities. This led the Soviet Union to develop orbital weapon systems like FOBS or to test ASAT devices, at the same time, making constant reference to the peaceful use of space. This imperative was, however, important enough to open the way to the conclusion in 1967 of the OST, which aimed at promoting "international co-operation in the peaceful exploration and use of outer space" (article XI).

The current situation shows that dual characteristics of military and civilian utilizations of space remain the same. Since the 1980s, military interest in space has greatly increased: The DOD budget surpassed the NASA budget in 1981, whereas other nations, such as China and France, have also become involved in military earth observations and communications programs. In July 1989 General M. Moiseev, chief of the Soviet general staff, declared that Soviet military space activities will help to enhance by up 100 percent the combat efficiency of the Soviet armed forces[3].

Military Satellites

Currently military satellites represent more than 50 percent of the launches and belong exclusively to the United States or the USSR. (see Fig. 4.1). According to the experts,[4] military satellites may be divided into five groups: reconnaissance satellites, communication satellites, navigation satellites, weather satellites, geodesic and scientific satellites.

1. *Reconnaissance and surveillance* satellites are used to provide secret information from space. They may be divided into four categories: "photographic" and radar-imaging intelligence, electronic intelligence, early warning, and ocean surveillance satellites.

---

[3] As quoted by *Jane's Defense Weekly*, 8 Dec. 1989.

[4] R. Banks, Documents de l'Assemblée de l'Atlantique Nord, Commission scientifique et technique, Rapport spécial sur l'exploitation de l'espace, février 1988. *SIPRI Yearbook 1989*, World Armaments and Disarmament (New York: Oxford University Press 1989).

Fig. 4.1.a Military Telecommunications and Remote-Sensing Systems.

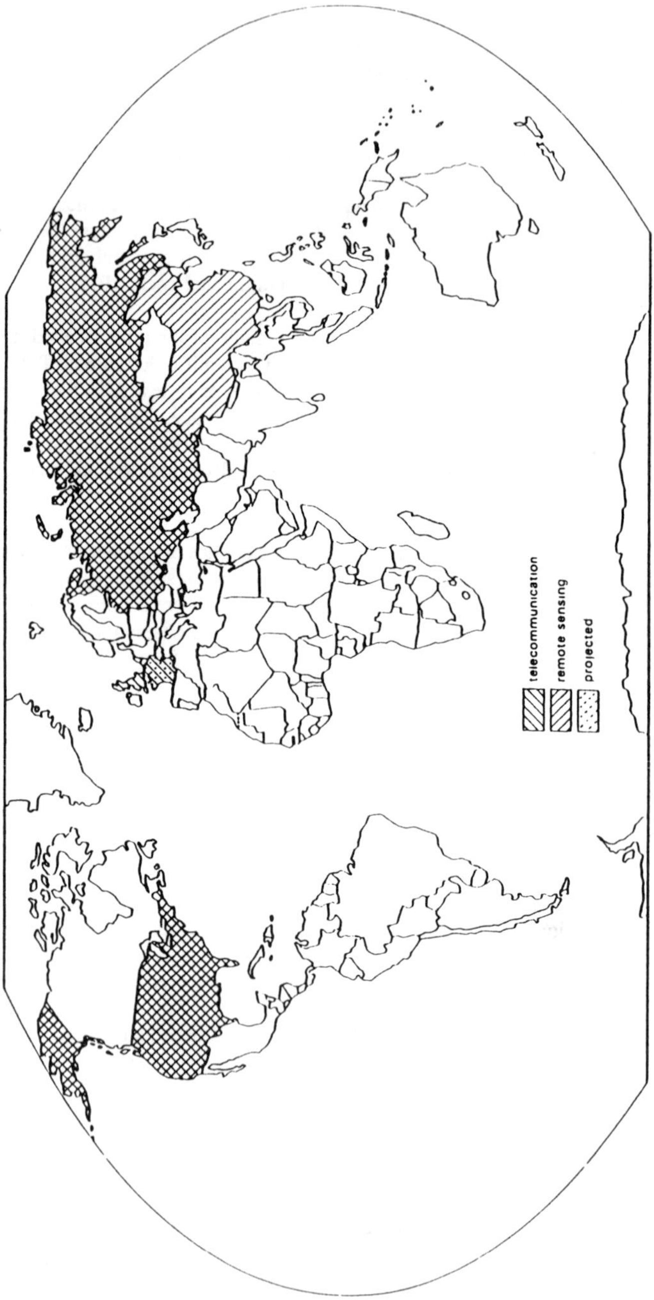

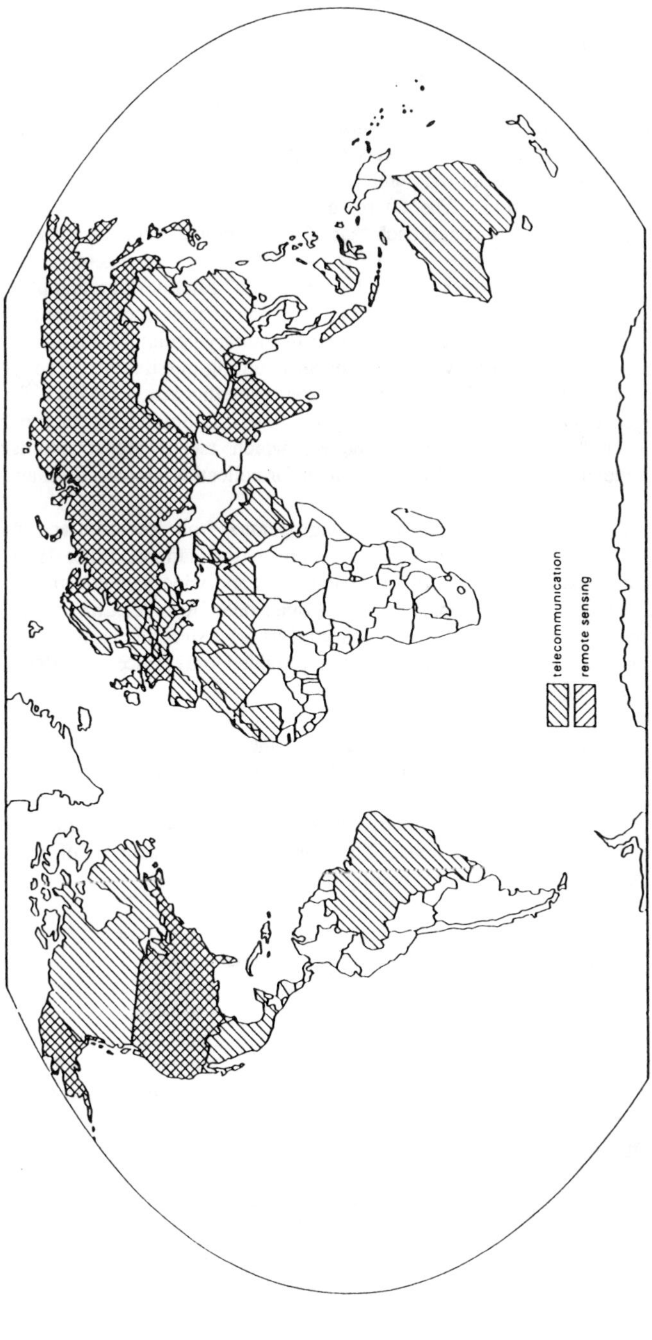

Fig. 4.1.b Civilian Telecommunications and Remote-Sensing Systems.

Reconnaissance satellites gather precise information at high resolution, particularly during a crisis, whereas surveillance systems carry out more regular monitoring of earth activities. The two functions are now generally done by the same satellite with various optical, infrared, and imaging sensors. These satellites have to transmit huge amounts of information. The Soviet third and fourth generations of Cosmos satellites use a close-look, film return system; the fifth generation uses electronic transmission to return images in near real time. The U.S. reconnaissance satellites Big Bird and KH-8 also used re-entry canisters, whereas KH-11s transmit data and KH-12 combines high resolution of film with real time electronic transmission. The Lacrosse system is the first *imaging radar* satellite whose resolution is supposed to approach that of photographic reconnaissance satellite.

Electronic intelligence, also called "Ferret" satellites, are complementary to remote-sensing satellites for collecting global information. Ferret satellites are specialized to pick up a variety of electronic transmissions used in military communications, detect frequencies of radars and collect telemetry data, for instance, during missile tests. The terms used to define these applications are ELINT for electronic intelligence, COMINT for communications intelligence, and TELINT for telemetry intelligence. Sometimes, they may all be called by the general term of SIGINT satellites, which means signal intelligence. They are complementary to Soviet ELINT satellites for collecting global information.

Soviet ELINT satellites have long operated from low orbit and transmit intermittent information when they are in sight of data reception stations. Since 1988, a new generation of geostationary satellites permits a continuous and wide area coverage. The U.S. strategy of ELINT capability is opposite to the Soviet one. Ryolite, Chalet, and Magnum systems have operated from GSO (Geostationary orbit) orbit since the early 1970s, and more precise coverage of northern regions has been provided only since mid-1980 by two Jumpseat satellites in highly elliptical orbits.

Early warning satellites provide permanent detection of both nuclear tests and missile launches. The U.S. system consists of five Defense Support Program satellites located on GSO orbit. They use huge telescope and infrared detectors. Their vulnerability in case of laser attacks is reduced by upgraded sensors and a new capability to communicate with any of the system's ground stations. The Soviet equivalent consists of a constellation of nine satellites in Molniya-type highly elliptical orbit.

Ocean surveillance satellites are used to locate and identify naval units. The Soviet Union operates EORSAT (Electronic Ocean Reconnaissance Satellites) and RORSAT (Radar Ocean Reconnaissance Satellites). As their acronyms indicate, the first system picks up transmissions (radio and radar), and the radar of the RORSAT system detects surface ships. The electrical power for the radar is generated by a nuclear reactor. Both are on LEO (Low Earth Orbit) orbit at, respectively, 400 km and 250 km. In contrast, the United States developed the Naval Ocean Surveillance System composed of a constellation of primary and secondary satellites launched into low polar orbit, picking up electronic transmission and locating the naval units by using triangulation.

2. *Communications* satellites are the second main family of military satellites (see Fig. 4.1a). They facilitate strategic and tactical communications between the command authority

and its units. They play a very important role on the battlefield or during naval or air attack missions. The U.S. Army, the Navy, and the Air Force use specific systems: DSCS (Defense Satellite Communications System), Fleetsatcom (Fleet Satellite Communications), and AFsatcom (Air Force Satellite Communications) whose transponders are carried on board of DSCS or Fleetsatcom satellites.

If one considers the location of these satellites in space, one will notice that most of the U.S. communications satellites, in particular the Fleetsatcom or the DSCS, are on the GSO orbit. The system is complemented by SDS (Satellite Data System) satellites in elliptical orbits. Their mission is to deliver in near real time information gathered by remote-sensing satellites like KH-11.

The Soviet's equivalent uses a specific orbit that is more elliptic and known by the civilian satellites that also are on this orbit, the "Molniya." The first, second, and third generation of Cosmos operate in low altitude orbits and the constellation is operating in different orbital planes. Telecommunication devices are also incorporated in geostationary satellites like the Raduga.

3. The Soviet and United States *navigation* and positioning satellites are very similar. The systems are formed by semisynchronous satellites (altitude of 20 000 km and period of 12 hours) and a low altitude constellation of small satellites. The U.S. Navstar Global Positioning System (GPS) will provide data to both military and civilian users, but the level of accuracy will be lower for the non-military customers. The development of the Soviet Glonass began after the U.S. GPS, it appears to be less advanced technologically.

4. *Weather* satellites provide information of utmost interest for the operations planning, knowledge of meteorological conditions, and their impacts on radar emission or for photography by reconnaissance satellites. The Soviet forces do not own specific military satellites and use the civilian low altitude Meteor-2 and -3 satellites. For their part, the U.S. forces use also civilian NOAA satellites and the air force owns a specific constellation of DMSP (Defense Meteorological Support Program) satellites.

5. *Geodetic* satellites provide precise measurements of the earth's shape. They are launched into low earth orbits and are essential for the guidance of missiles. Scientific satellites complete the panorama of military satellites and study outer space elements like magnetosphere or Van Allen belts.

This general survey of military space programs must not forget manned missions. The Soviet space stations Salyut in the past and Mir today allow monitoring the earth from space and carrying out scientific research. The U.S. Space Shuttle is also used for military purposes even if they are still sporadic. The development of Freedom station will probably lead to a permanent military presence in space.

Military uses of space offer a wide variety of missions: Soviet and American satellites are the most numerous, but China has also developed some low earth orbit and GSO satellites for reconnaissance, weather, and communications missions. The United Kingdom continues the Skynet telecommunications program and France continues to develop Syracuse system on

Telecom 1 and 2 satellites. Israël seems to be interested in an independent reconnaissance capability as indicated by the launches of Offeq satellites.

Military satellites perform different tasks, which do not appear in themselves particularly peaceful or non-peaceful. A discussion of civilian satellites is useful to see if some activities in space are strictly military or not.

Civilian Satellites

In the field of earth observation, a considerable amount of data on meteorological and remote sensing is available. International cooperation is growing in the field of the world meteorological surveillance (see Fig. 4.1b). It is now done by different national satellites belonging to: Europe (METEOSAT-3 and 4), India (INSAT), Japan (GMS-3), and the United States (GOES EAST and GOES WEST). The Cosmos 1940, launched in 1988 and sometimes called GOMS, also participates in the Meteorological Watch. This network is complemented by low earth orbit systems like the U.S. NOAA and the Soviet Meteor. Data are openly provided and can be bought by users.

Weather satellites are more closely related to civilian spacecraft than military ones. The Soviet Meteor, which performs both civilian and military needs, and the U.S. DMSP are quite similar to NOAA civilian satellites.

Remote-sensing data are less numerous and more expensive to acquire; the commercialization of the U.S. Landsat, and the French Spot images have increased acquisition of data on a non-discrimination basis. Japanese MOS (Maritime Observation Satellite) or Indian IRS (Indian Remote Sensing Satellite) data are not commercialized yet, but earth's observation, meteorological, and remote-sensing civilian data are available to everybody. The future European radar system ERS (Earth Remote Sensing), the Japanese JERS, and the Canadian Radarsat will increase the capabilities of the existing civilian satellites to all weather, day, and night observations.

Early warning satellites have no civilian counterpart, but ocean observation satellites have existed and others will be launched soon (ERS, Topex-Poseidon) even if they perform more scientific missions than the fleet search done by military surveillance satellites.

As far as telecommunications are concerned, there has been a tremendous increase in information transfer allowed by the development of international, regional, or domestic systems at the same cost whatever the profit of specific channels devoted to specific geographical areas may be. This provided Third World countries a better chance for integration in the world economy. Thus international Intelsat permits transmission of information all around the world. Regional satellites systems such as ARABSAT or EUTELSAT, and domestic programs such as AUSSAT, BRASILSAT, or TELECOM, perform the same role in helping to promote economic exchanges.

New tools, such as those provided by navigation satellites also become more important for civilian maritime activities using specific channels of the GPS system and probably soon for every mobile transportation system with specific civilian programs such as the European Locstar. Finally, various scientific satellites perform research, for example, about earth, solar environment, and astronomy. Their results are opened to the scientific community and very often correspond to wide cooperation.

Thus one can underline that the satellites concerned with peaceful or non-peaceful missions in space are numerous and offer a wide variety of functions both in military and civilian fields. Figure 4.1.b shows the geographical location of telecommunications and remote-sensing programs.

In summary, one has to consider whether peaceful or non-peaceful purposes are more obvious according to geographical or functional criteria applied to study either military or civilian satellites activities.

The geographical criteria are based on the orbit on which satellites circulate. According to the altitude of the satellite, two main orbits are the Low Earth Orbit (LEO) and the geostationary (GSO).

Functional criteria are divided into categories such as remote-sensing activities (including meteorology, photographic reconnaissance, ocean observation, and early warning), telecommunications, electronic intelligence, and navigation. Neither geographical nor functional criteria guarantee distinction between the peaceful or non-peaceful nature of satellites.

The differentiation between civilian and military activities may result from the identification of the owner and the user of a satellite. For example, in the United States, the financing and operating of a program by NASA or the Department of Defense may generally reveal the civilian or military nature of the activity. The organization of Soviet space programs, however, doesn't allow the same analyses. In fact military and civilian programs are not differentiated and the creation of the Soyouzkarta firm shows that the distinction between commercial satellites pictures and military data may be blurred.[5]

The differentiation may also result from the performance and the type of equipment carried by the satellite. Even in this case, the dividing line between peaceful and non-peaceful activities is not that easy. What may matter most is the context in which a country will use data collected or transmitted by a satellite. The launching by a neutral country of an ocean surveillance satellite may appear more "peaceful" than the launching of the same type of vehicle by a country possessing an ocean-going navy spread on all the seas, as is the case with the superpowers. Let us look, first, at peaceful uses of both military and civilian satellites, then at non-peaceful uses, and, in conclusion, at a general thought to preserve peace in space.

---

[5] Soyouzkarta sells selected pictures with 5, 6 m of resolution of sites that may have been observed for intelligence purposes. The military origin of these photographs is enhanced by the locations of the available data, which coïncide with many military installations, and the refusal of Soyouzkarta to sell to Western countries pictures of the Soviet Union and their allies.

II - What about Peaceful Uses of Satellites?

Soviet and U.S. Military Space Programs

Only the United States and the Soviet Union have a complete and permanent military system of information gathered from space. Photographic reconnaissance satellites have been used since 1960 and were frequently launched (Fig. 4.2), especially by the Soviet Union whose strategy in space is very different from the U.S. one, mainly because for strategic and technical reasons. In a typical year, the Soviet Union will conduct five times more launches than the United States, largely because of the less complexity of the spacecraft, the sensors, and the ways of transmission, which make their duration shorter. In contrast, the United States relies heavily on small numbers of highly sophisticated durable satellites. The Soviet Union currently uses different generations of photographic reconnaissance satellites usually divided according to their type of resolution and lifetime. In 1988, 32 Kosmos photographic satellites were launched to satisfy military as well as civilian needs. Three major classes are considered. The satellites belonging to the so-called third-generation (Vostok type) usually disappear from their orbit in a two-week period. The fourth-generation satellites (Soyuz type) can remain in orbit for two months, and the fifth generation is able to orbit for at least six months. The orbital characteristics as well as, of course, resolution of each generation are different (altitude, eccentricity, inclination). A new generation may have been born in 1989 with the launch of Kosmos 2031.

In comparison, the United States has deployed only two-generation "Key Hole" satellites, but their lifetime is longer from three years for KH-11 up to probably six years for KH-12. The launching failures after the Challenger accident restricted the U.S. photo-reconnaissance capability to a single satellite during 1986 and a large part of 1987. In 1989 three satellites were in orbit and provided high and medium resolution capabilities. They do not eject photographs in canisters, but instead convert images into digital electronic signals and transmit them via microwave links to ground stations where they are reconstructed into photographic images.

The peaceful aspects of these programs depend on the better provision of more accurate and permanent information. The low or medium resolution systems observe large segments of a territory in proportion to the orbit inclination provided itself by the latitude of the launching site. They allow global monitoring of the earth when the different orbits include polar satellites. Depending on the results of information received, or on particular events occurring in the world, "close look" satellites are used to focus on "hot spots." These very high resolution data are vital for a good evaluation of the situation during crisis. Their non-intrusive nature is a great advantage to appreciate more objectively the strategy of other countries. In this sense the help of SIGINT satellites for intelligence gathering is essential. Both the United States and the Soviet Union have developed complementary constellations in circular and elliptic orbits, even if U.S. satellites are more numerous in GSO orbit (see Fig. 4.3a, b).

Fig. 4.2 Military Reconnaissance Satellites.

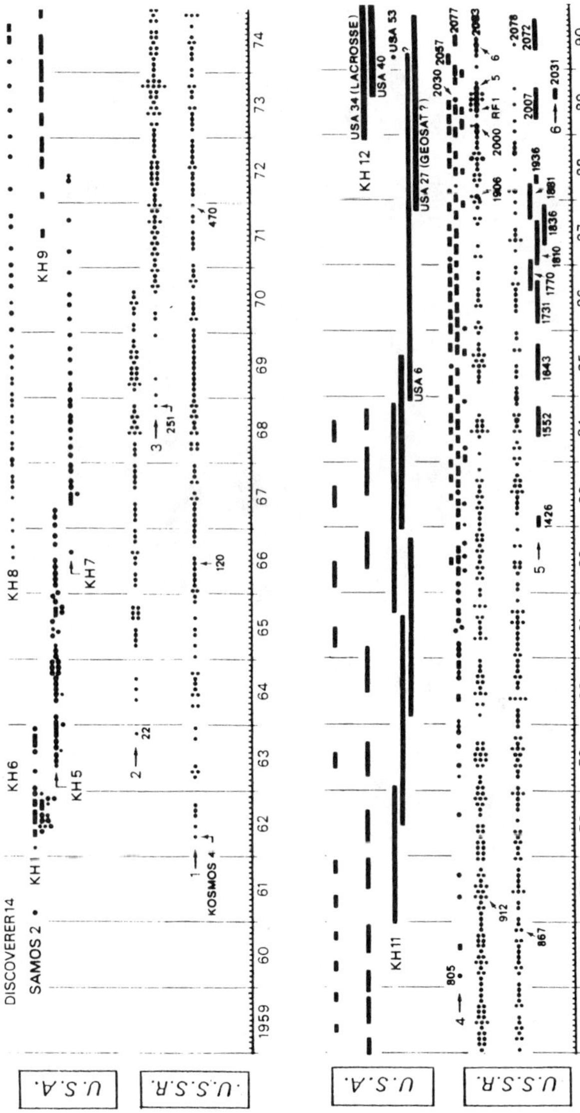

Source: F. Verger, La vie des Sciences, mars-avril 1989 : L'Intérêt d'observer la Terre depuis l'espace.

Fig 4. 3a Usual soviet satellite orbits

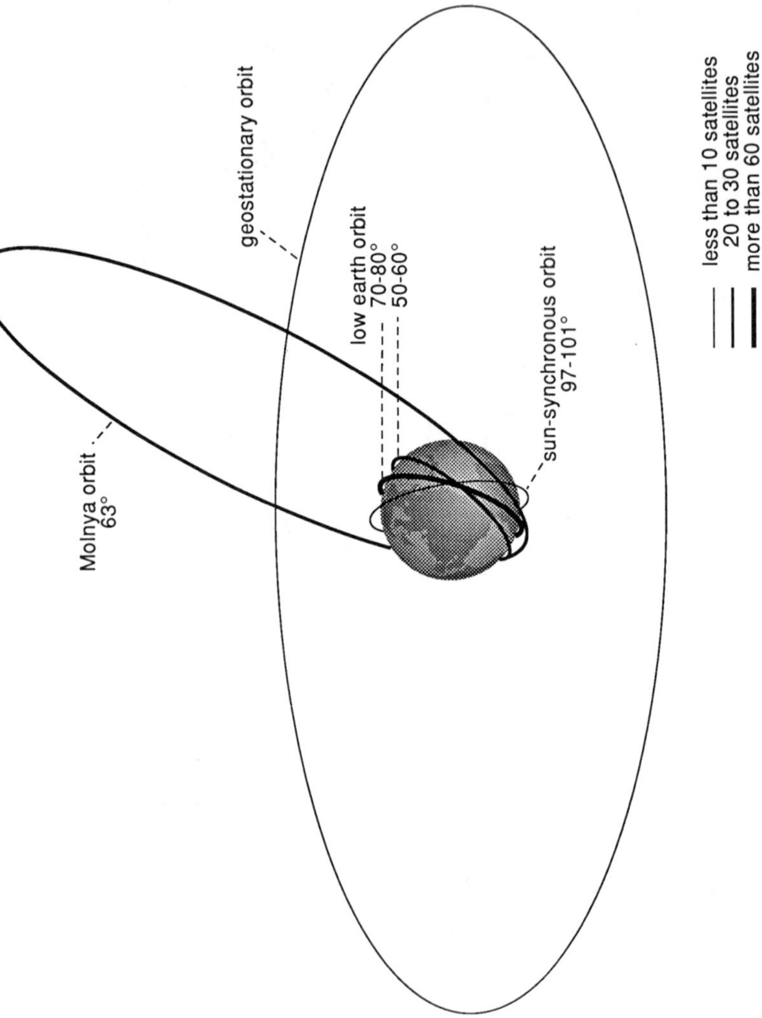

Fig 4. 3b Usual U.S. satellite orbits

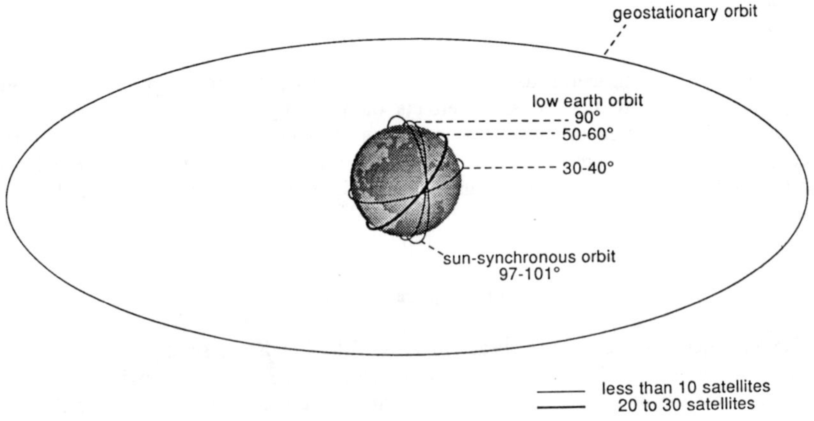

Fig 4. 3c Usual satellite orbits of other countries

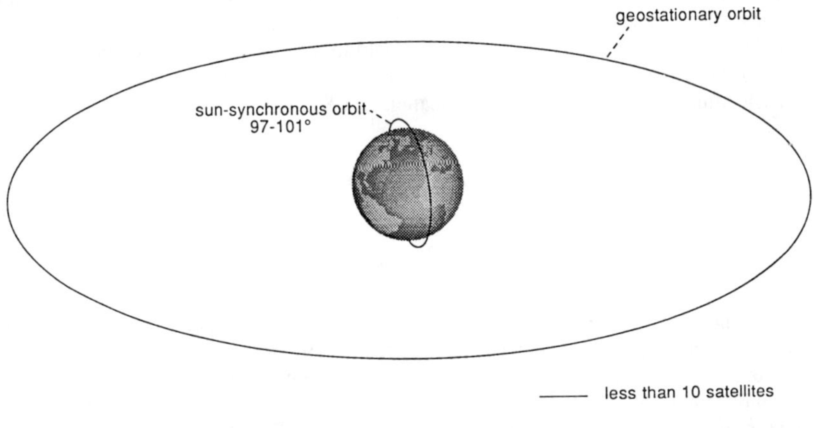

The "peaceful" value of these satellites is noticeable especially during crisis management. They may provide information about the real status of an opponent forces. In some circumstances they can contribute to avoiding an escalation during a crisis.

Thus military satellites may be helpful during a phase of tension between two states or a group of states belonging to opposing alliances. They are also now becoming an essential tool that permits verification of compliance with arms control agreements by NTMs or national technical means.

Therefore the peaceful uses of space are related to the acquisition of global information. This may have positive effects on reducing the threat of escalation, on increasing stability and favoring the application of arms control agreements. A deterrent effect may also be considered when one knows that a better understanding of the behavior of an opponent may reduce the risk of misunderstandings and misperceptions should a crisis occur.

Civilian Space Programs

Civilian satellites have, of course, peaceful uses. The Landsat system was built "for the benefit of all mankind." The American Landsat, the French Spot, or the Soviet Priroda data are essential for the better management of a territory, the prevention of natural risks, crop surveillance, and urban development. All these uses are, no doubt, peaceful, but the increased level of resolution of these systems brings them closer to military systems. In the same way communications satellites like the Soviet Ekran, Molniya, and Raduga, the European Eutelsat, such as the U.S. Comsat or DBS satellites, the Japanese CS-3, and the Indian Insat perform peaceful tasks as integration in national and international exchange services of telephone, television, or education. However, the increased capacity of civilian satellites and the high technology they develop have complicated the analysis of their uses. Civilian communications satellites sometimes carry on board military systems, as seen in France with Syracuse used by the military and carried on board a telecommunication satellite operated by the French Ministry of Post, Telecommunication and Space.

III - What about Non-Peaceful Uses?

Military Satellites

The ASAT program is one example of "offensive" spacecraft launched into orbit. Initially developed by the Soviet Union and the United States, both countries have enacted a moratorium on their ASAT activities. Any ASAT program is clearly a destabilizing system for it is threatening satellites with positive aspects that have already been stressed, especially in the case of crisis management and greater transparency. The prohibition of an ASAT race would enable the superpowers to publicize a more peaceful image of their military activities in space.

The development of the SDI (Strategic Defense Initiative) program raised other issues such as the interpretations to be given to the 1972 ABM Treaty in the field of delimitation between "research" and "development," definitions of components (missiles, launchers, and radars), and dual character of technology.

The fact remains that this strategic approach to the uses of space is stirring up the space arms race. However, apart from these systems, some categories of military satellites used in peacetime for stabilizing purposes may be used as efficient military tools if needed. A good example of this kind is given by military meteorological satellites, one of the first systems developed by the U.S. military forces. Two Defense Meteorological Satellite Program satellites (DMSP) are currently orbiting in a sun-synchronous orbit and provide meteorological information four times a day on a special area. They analyze cloud patterns, measure precipitation, surface wind speed, and even electron distribution in the upper atmosphere. They also provide the information to investigate radar calibration and efficient bands of radio propagation in high frequency.

In wartime, those data are vital to air operations. They help determine whether air activities are possible, which route aircrafts must follow, and at which altitude they can navigate. Naval forces also use meteorological data of high precision to conduct their operations.

The Soviet Armed Forces do not own specific weather satellites, but they do use civilian meteorological satellites, such as Meteor-2, for identical purposes.

In the same vein, the GPS and Glonass systems, also used for civilian activities, are particularly valuable for many military operations. Thus the need for accurate navigation data and localization is vital during war operations. First to be developed, the U.S. Transit program, operational in the mid-1960s, was able to provide navigation data for slow mobile systems such as submarines. The four experimental Navstar satellites, launched in 1978, opened a new era under the direction of the Air Force Division. Navstar is currently operational and a 12-satellite constellation is able to offer a precise location to within 15 m and to evaluate the velocity to within 0,1 m per second. A two dimension is planned for in 1990, whereas worldwide three-dimensional capability from an 18-satellite constellation is expected in 1991.

The Soviet navigation and positioning system Glonass is very similar to Navstar even if the level of accuracy seems less (estimated to 100 m). The geodesic satellites are clearly associated with navigation satellites for accurate targeting by long-range tactical and strategic weapons. Both the United States and the Soviet Union have developed geodesic programs such as, respectively, Lageos and Etalon.

If weather, navigation, and geodetic satellites are of the utmost importance for air, naval, and nuclear operations in wartime, reconnaissance and ELINT satellites play a decisive role of force multipliers in ground and nuclear operations.

The United States and the Soviet Union have chosen different policies for reconnaissance systems. In the military field, photographic spacecraft, using sensitive optical and infrared films, can monitor the deployment of forces and troop concentrations, look after missile silos, shipyards fleets deployments, and also follow development of new bases. The reconnaissance satellites have definitely affected war planning: surprise attack is, at least, greatly reduced. They provide invaluable data for precise targeting against cities, depots, airfields, harbors, and all military targets. They also produce mapping of enemy territory, which is very helpful for attacking. During conflicts, the assessment of the order of battle adopted by an opponent, the localization of concentration of forces, the tactical movements of military assessment should offer a major superiority, but the problem of data transmission is not resolved in particular for remote areas. The U.S. military is currently developing a comprehensive network of satellites and they are more advanced compared with those of the Soviet Union. Today, commanders can receive information on the situation on the battlefield and new generation satellites that transmit data directly in real time to commanders are expected for the year 2000. However, high resolution monitoring of events appears almost impossible in real time for reasons of orbital mechanics.

If the U.S. KH and the Soviet Kosmos seem very useful for war planning, they also play a crucial role during crisis management. Electronic reconnaissance satellites play a similar part in wartime. Data gathered by radar and radio-emitters located in ground, air or sea, even from mobile units, allow management of an electronic order of battle. They also permit the development of countermeasures by listening to radar signals and providing early warning of movements.

Ocean reconnaissance satellites are more dedicated to naval operations. The Soviet Rorsat system does not appear truly operational, but in the future it will represent a real threat to U.S. naval forces by reporting, with great accuracy, the movements of such forces. In addition, the use of nuclear system in Rorsat satellites raises serious discussion. In fact, the use of reactors in space is not new. In 1965 the United States launched the first space reactor and different studies have been conduct during the 1970s. The Strategic Defense Initiative has given a new impetus to the idea of nuclear reactors for use, for instance, with DEW satellites (Directed Energy Weapons), and the SP 100 project has received important fundings. In 1978 President Carter proposed a ban on orbiting reactors, and the idea has to be seriously considered for environmental and arms control reasons. In fact, the failure of two Rorsat satellites has clearly showed the risk of radioactive contamination, especially from their low altitudes. The usual procedure in the Rorsat system is to place the used reactors in higher orbits (950 km), which increases the space debris pollution. Finally, radiations from the Rorsat in space interfere with gamma ray astronomical observations. Moreover, the strategical impact of Rorsat satellites makes them priority targets for the U.S. ASAT capability. In that sense, Rorsat satellites may appear as non-peaceful satellites. However, ELINT satellites provide similar information so that they would appear to qualify as non-peaceful satellites also.

The most useful satellites, especially in wartime, remain telecommunication satellites. Command, Control, and Communication Intelligence (C^3I) appear as essential elements of modern warfare. The proliferation, diversity, and redundancy of communications satellites are necessary in order to guarantee the security of the system.

The U.S. military communications uses, for example, different systems such as DSCS, Fleetsatcom, SDS, Leasat, and specific transponders known as Afsatcom carried on board various satellites. They are located in geostationary orbit and utilize ultrahigh-frequency (UHF) or super high-frequency (SHF) bands. The Soviet military communications satellites use high frequency very close to the civilian system. They are divided in two systems: two low altitude Kosmos constellation in circular orbit, Molniya satellites with highly elliptical orbit and Geosynchronous satellites.

Civilian Satellites

Civilian communications satellites may also be used for military communications and its share should increase in case of crisis.

Thus in wartime many satellites will extensively be used for non-peaceful purposes; they will also play a role of force multipliers.

Even when the utilization of a satellite is ruled by limits and legislations such as the telecommunication satellites, it is increasingly difficult to differentiate between military or civilian uses. For instance, the attribution of a location in the GSO corresponds to specific requirements made by a country for a particular activity utilizing a precise frequency band. The agreement does not imply a restriction regarding the function of the satellite being either military or civilian. Moreover, it is not required to mention the mission of the satellite. Also, it is possible to utilize bands that have not been officially attributed. The only limits are to avoid jamming another satellite that is already officially declared to the IUT.

For the most part, the Soviet Union and the United States have shared the monopoly of earth military high resolution monitoring. Currently the civilian systems first developed in the communications field, then in meteorology and remote sensing, may also provide to other countries strategical data. The sale of Landsat and SPOT images, the reception of Meteosat or Goes meteorological pictures, the leasing of communications channels on Intelsat or other similar spacecraft offer quite similar services even if the accuracy level is less than the military one.

So, technological progress will provide more crucial questions on the general use of space, becoming the ground of a sharp political, economic, and military race.

The increased sophistication of civilian satellites will soon provide them with characteristics very close to those needed for military applications. With 10-m resolution in one panchromatic band and stereoscopic capability, SPOT offers to its buyers "decision assistance concerning a target located deep inside a zone inaccessible to reconnaissance planes..." and "penetration assistance." It may thus appear that it will be increasingly difficult to draw a clear distinction regarding the purpose of their use. Consequently as far as technology contributes to blur the distinction between military and civilian capabilities, the notion of peaceful and non-peaceful uses of space may appear increasingly irrelevant if one continues to rely on criteria traditionally employed to differentiate military and civilian applications.

IV - Conclusions

Since 1957 more than 3,000 satellites have been launched. They are of different types, either autonomous unmanned vehicles or manned flights. They perform various functions, among which the most significant ones are those for scientific research, earth observation, and telecommunications.

The civilian and military programs are equally emphasized and there is no clear way to differentiate between them. Certain criteria allow identification of military functions (type of launcher, altitude, lifetime, sensors used, types of documents produced, frequency bands, or non-accessibility to data), but that does not automatically imply that it resorts to non-peaceful activities. If one assumes that the opposite of peaceful is warlike, a passive military use of space (observation, surveillance, or communication) is not inconsistent with the principles enacted by the United Nations. Consequently, the real issue is the aggressive use of space (attack or fight).

Actually, the notion of non-peaceful use of space seems to cover without any ambiguity weapons systems positioned in space, either employed against targets located in earth or in space. In this case, one must take into account other international rules: interdiction of mass destruction weapons (OST), prohibition of the use of force in international relations (as stated in the Charter of the United Nations), or rules enacted in several bilateral treaties (ABM or NTM).

Bibliography

de la Rochère, Jacqueline Dutheil (ed), *Droit de l'Espace*, Paris: Editions Pedone, 1988.

Jane's Spaceflight Directory 1988-1989, London: Jane's Publishing Co., 1989, 644p. SIPRI Yearbook 1989, World Armaments and Disarmament, New York: Oxford University Press, 1989.

B. Jasani and T. Sakata (eds), *Satellites for Arms Control and Crisis Monitoring*, New York: Oxford University Press, 1987.

Johnson, Nicholas L., *Soviet Military Strategy in Space*, London: Jane's Publishing Co., 1987.

Krepon, M., Spying from Space, *Foreign Policy* (summer 1989).

Primack, J.R., and al. Space Reactor Arms Control, *Science and Global Security*, 1 (1989), 49-72.

Soviet Space Programs: 1981-87, Committee on Commerce Science, and Transportation, Part 2, U.S. Senate, April 1989.

Stares, Paul B., *Space and National Security*, Washington DC: The Brookings Institutions, 1987.

Verger, F., *L'intérêt d'observer la Terre depuis l'espace*, la *Vie des Sciences*, tome 6, No. 2, pp. 93-112, Paris, Gauthier Villars, 1989.

Zimmerman, P.D., *Remote Sensing Satellites, Superpower Relations and Public Diplomacy*, New York: St. Martin's Press, 1990.

UNIDIR. *Disarmament, Problems Related to Outer Space*, New York: United Nations, 1988.

CHAPTER 5: PROBLEMS OF DEFINITION: A VIEW OF AN EMERGING SPACE POWER

S. Chandrashekar

The absence of a clear international regime governing the "peaceful" and "non-peaceful" uses of outer space poses major contradictions between domestic and international policies to an emerging space power. Coupled with the lack of progress in achieving genuine "peaceful" uses of space, this will eventually lead to more countries using space for military purposes. The word "peaceful" has two essential components associated with it. These are an absence of force or conflict and the presence of calm and tranquillity. The current international regime of space is not a peaceful regime. Activities related to defensive weapons and support military functions can be carried out in outer space.

A categorization of the different uses of space along with the legal regimes associated with them indicated that activities related to the weaponization of space are a major area of concern requiring immediate initiatives. Regulation of support military activities, although necessary, could be taken up later as phase two.

Different approaches are possible toward the elimination of space weapons. In one approach a generic ban could be imposed. This would logically require a definition of space. In another approach, a ban on ASAT and BMD weapons only could be imposed. This would involve definition of "an object in space." To categorize objects in space, the orbit characteristic of a satellite, and a "cut-off velocity," and "time of flight" criteria for a ballistic missile could be used. Other variants of these basic approaches could also be considered.

An international verification mechanism that could include tracking of space objects, monitoring telemetry, and independent observation satellites, could monitor non-peaceful activity. Parameters such as radiation hardening, weight, power, nature of telemetry transmission, free availability of data and satellite services, and international participation could be used as additional elements to categorize peaceful uses.

The world "peaceful" can no doubt be defined to facilitate progress in the peaceful uses of outer space, but the approach suggested in this Chapter involves the clear and unambiguous definition of the term "non-peaceful."

The major elements of such a definition should include:

- The use of force or the threat of use of force by or against a space object.
- The use of a space weapon.
- The use of space objects to aid and assist in military operations.

I - Background

The question of peace in space and the peaceful uses of outer space has been a major issue before the international community. Several excellent reviews have been published.[1]

So long as there were only two players in the space race, geopolitical interests overshadowed other considerations and military uses emerged as a major requirement. Early warning, reconnaissance, navigation, and military communications were the early drivers of the military effort. The first nuclear tests in space took place in 1962. It is reasonable to conclude that the resulting damage to several satellites was at least partly responsible for the prohibition of nuclear explosions in outer space as enunciated in the Partial Test Ban Treaty and reemphasized later in the Outer Space Treaty (OST) of 1967.[2]

At the United Nations consensus was finally achieved on the contents and text of a treaty on the activities of states in outer space, which came into force in 1967.[3]

The late 1960s and the early 1970s witnessed new developments, including new weapon systems for use in and against objects in space and on earth.

The question of arms in space and the peaceful and non-peaceful uses of outer space became particularly important after the initiation of SDI in 1983.[4] Since then space militarization, space weaponization, peaceful and non-peaceful uses of outer space, and the extension of the arms race into space have all become terms much used, discussed, and elaborated upon in several international fora.

However, in spite of the rhetoric and the plethora of viewpoints on approaches, little progress has been achieved in realizing the ideal goal of "the exploration and use of outer space for peaceful purposes," which is enshrined in the preamble to OST.

Recent progress in arms limitations talks between the superpowers has re-kindled hopes that progress on the ticklish issues of peaceful/non-peaceful uses of outer space is

---

[1] Ivan A. Vlasic, "Disarmament Decade, Outer Space and International Law," *Mc Gill Law Journal*, Vol. 26, no. 2, 1981; William J. Durch, *National Interests and the Military Uses of Space* (Cambridge: Ballinger, 1986); Yu. M. Kolossov and S.G. Stashevskii, "The struggle for a peaceful outer space -- Legal issues," translation from the original Russian by Dr. Sheila Iyer, Publication of the ISRO Satellite Center, Bangalore, India, 1986; UNIDIR, "Disarmament: Problems Related to Outer Space," UN publication, 1987.

[2] Treaty Banning Nuclear Weapons Tests in the Atmosphere, in Outer Space and Under Water, 1963; Clause G. Goetzel et al, *Space Materials Handbook* (Reading, MA: Addision Wesley, 1965), pp. 56 - 61, 481 - 86.

[3] "Treaty on Principles Governing the Activities of States in the Exploration and Use of Outer Space, including the Moon and other Celestial Bodies," UN Resolution 2222 (xxi) and annex 3 Dec. 1966.

[4] President Reagan's speech on defense spending and defensive technology -- Administration of Ronald Reagan, 23 March, 1983 from *Weekly Compilation of Presidential Documents*, 19, no. 12 (Mar. 28, 1983), 423 - 66.

indeed possible. The purpose of this Chapter is to delve a little deeper into the larger issues of war and peace in space and emerge with views, definitions, and approaches that could become useful in case progress on these difficult issues does indeed take place. The viewpoint is that of an emerging space power that has realized certain capabilities in the peaceful uses of outer space.

II - The Typical Emerging Space Power

The emerging space power pictured here is a fairly big country with a fairly large pool of trained personnel and reasonable material and natural resources. On the political front, it has aspirations toward achieving or helping to achieve, a lasting peaceful international system in which cooperation and not confrontation is the normal mode of working. Global disarmament, a more equitable distribution of wealth, and a world order that operates on a system of mutual respect for different cultures, ideologies, and socio-political systems are the main pillars of its international policy.

On the domestic front it has a large pool of technical talent and attendant problems of employment, poverty, and population growth. As a medium power in its region, it has a fairly big military setup. Relationships with its neighbors are conditioned to a large extent by its potential to become a major economic, political, and military power in the region as well as by historical factors.

The reconciliation of its international policy with its regional obligations, under the framework of superpower or East-West differences, very often results in domestic and international pressures that make the development route chosen by the country difficult to follow. Technology sanctions, embargoes under the apparent guise of non-proliferation of potential military technologies are major factors to contend with as a part of this international order.[5]

Within the country, the emerging space power is often faced with extremely difficult choices. There is pressure from the military establishment to plough more resources into military programs. At the same time, development requirements for economic growth also need large resources. At some stage the country's international role as peacemaker comes under fire because of domestic pressures brought about by regional compulsions. A push toward military strength in a world order that respects only strength is a part of this evolution.

A country that is at this crossroads of policy choices and that has a space program that is civilian, now has to contend with a push toward the military uses of space. The program that began as a judicious blend of indigenous development and international cooperation suddenly runs into rough weather.

The emerging space power looks around and finds an international situation in space that is chaotic. The international regime of space (with few exceptions) permits almost any military use of space. However, for the emerging space power, the realization of even

[5] "Missile Technology Control Regime," announcement by the White House, 16 April, 1987.

marginal capabilities in launcher and satellite design and construction evokes considerable protests from developed countries who want to preserve their military domination and use of space.[6] Because these apprehensions on the part of the advanced space powers are very often not coupled with any concrete disarmament measures, these protests and sanctions only strengthen the determination of the emerging space power to become independent in terms of the capability to use space.

A reasonably clear, unambiguous international order on the uses of space, specifically on the peaceful uses of outer space, is necessary to resolve many of these contradictions that an emerging space power encounters. The absence of such a system, and the lack of progress toward "disarmament in space," force all potential space powers to embark on a path similar to those taken by the more developed countries.

The central theme of this Chapter is how an emerging space power views problems related to peaceful/non-peaceful uses of outer space. It addresses in some detail the elements of what constitutes peaceful and non-peaceful uses and outlines specific areas of concern. It also attempts to provide a framework on how discussions and negotiations on the vital issues of war and peace in space can progress, keeping in mind real problems. It deals with a host of phrases, terms, and words such as "peaceful," "non-peaceful," "military," "space weapons," "arms race in space" in as simple a way as possible. Rather than dealing with semantics, the approach focuses on concepts and evolves definitions based on these concepts.

III - The Concept of Peaceful Use

The word "peaceful" as defined in Webster's dictionary lists the following meanings:[7]

- "peaceable," "untroubled by conflict agitation or commotion," "quiet," "tranquil," "of or relating to a state or time of peace," "devoid of violence or force."

The word "peaceable" has the following meanings:

- "disposed to peace," "quietly behaved," "marked by freedom from strife or disorder."

The thrust of these meanings seem to point towards two general concepts associated with the word "peaceful":

- "peaceful" involves the absence of conflict, agitation or commotion. It also involves the absence of violence or force.
- "peaceful" includes in its meaning the presence of "quiet," "tranquillity."

---

[6] Ibid.

[7] *Websters Seventh New Collegiate Dictionary* (Springfield, MA: G & C Merriam, 1965).

The absence of conflict and the presence of tranquillity are both ingredients in a "peaceful" use.

In the specific context of space, the word "peaceful" occurs in several articles of the Outer Space Treaty.[8] Paragraph two of the preamble recognizes "the common interest of all mankind in the progress of the exploration and use of outer space for peaceful purposes." Paragraph four expresses the desire "to contribute to broad international cooperation in the scientific as well as the legal aspects of the exploration and use of outer space for peaceful purposes." Paragraph two of Article IV reads "The Moon and other Celestial Bodies shall be used by all States Parties to the Treaty exclusively for peaceful purposes" and further elaborates on what military activities are forbidden on the moon and other celestial bodies. Paragraph one of Article IV bans nuclear weapons and weapons of mass destruction in space, but it does not explicitly state that outer space shall be used exclusively for peaceful purposes. Article IX stipulates consultation concerning activities or experiments that "would cause potentially harmful interference with activities of other States Parties in the peaceful exploration and use of outer space, including the Moon and other celestial bodies." Finally, Article XI talks about states agreeing to inform the secretary general of the UN as well as the public and the international scientific community about the nature, conduct, locations and results of space activities in order to promote international cooperation in the peaceful uses of outer space.

An analysis of this treaty indicates the following:

- The treaty is clear that the moon and other celestial bodies shall be used exclusively for peaceful purposes. All activities that are military in nature, including the testing of weapons, are prohibited on the moon and other celestial bodies.
- The treaty in no way binds states to use space exclusively for peaceful purposes. Non-aggressive and defensive military activities (which are permitted by international law and the UN Charter) are allowed in outer space.

The treaty wording is such that except for the stationing of nuclear weapons and weapons of mass destruction, all other forms of military activity are permitted in outer space. A comparison with paragraph 1 Article I of the Antarctic Treaty[9] makes this distinction clear: "Antarctica shall be used for peaceful purposes. There shall be prohibited interalia any measures of a military nature such as the establishment of military bases and fortifications, the carrying out of military maneuvers as well as the testing of any type of weapons." Nowhere in its text does the OST make any such similar declaration. Only certain kinds of activities related to nuclear weapons and weapons of mass destruction are prohibited. It is not binding on states parties to the treaty to use outer space only for peaceful purposes. It rather puts some limits on some uses. Military use of space is therefore legal.

---

[8] "Treaty on Principles Governing the Activities of States in the Exploration and Use of Outer Space, including the Moon and the Celestial Bodies", *op cit.*

[9] Antarctic Treaty, Multilateral Antarctic Treaty signed in Washington, DC, 1 Dec., 1959.

A number of divergent views have added confusion to the original intention of the treaty makers. Most of them interpret the term "peaceful" to suit a particular point of view. One school of thought equates the term "peaceful" with "non-aggressive."[10] Thus self-defense is a peaceful use in their opinion. To justify this, they refer to Article 51 of the UN Charter, which provides states with the right of self-defense against aggression.

Article 51 of the UN Charter reads.

> Nothing in the present Charter shall impair the inherent right of individual or collective self defense if an armed attack occurs against a Member of the United Nations, until the Security Council has taken measures necessary to maintain international peace and security. Measures taken by Members in the exercise of this right of self defense shall be immediately reported to the Security Council and shall not in any way affect the authority and responsibility of the Security Council under the present Charter to take at any time such action as it deems necessary in order to maintain or restore international peace and security.

A careful perusal of Article 51 reveals the fallacy in this argument. The distinction between an aggressive act or a preemptive defensive action is very difficult to make. In referring to Article 51, many sources forget to refer to the Security Council role to "maintain or restore international peace and security." This clearly indicates that the question of self- defense arises only when the peace has been violated or is likely to be violated.

This justification, however, is only of academic interest because the OST does not forbid the stationing of defensive weapons in space. Nor does it prohibit other kinds of support activities.

The Antarctic Treaty, which clearly stipulates that Antarctica shall be used only for peaceful purposes and which explicitly prohibits any measures of a military nature, does not invoke the UN Charter or the right of self-defense. If nations had desired a peaceful use of space, they could have come to a similar arrangement on space. It is clear that peaceful is not synonymous with "non-aggressive."[11]

There is another school of thought that considers non-weapon military use as a peaceful use.[12] This is, of course, a variation of the "non-aggressive" use being peaceful with the addition that weapon-related activities in space even for defensive purposes cannot be considered peaceful. This argument makes no sense, because in a fundamental way the use of space by the armed forces directly or indirectly for their operations aids or assists in the use or the threat of use of force. This therefore cannot be construed to be a peaceful activity.

---

[10] Statement of FRG Representative in Conference of Disarmament, Document CD/PV.185.

[11] M. Marcov, "Disarmament and Peaceful Purposes, Provisions in the 1967 Outer Space Treaty," *Journal of Space Law*, 4, No. 1 (1976), 7.

[12] R.L. Garwin, "Are We on the Verge of an Arms Race in Space?" *Bulletin of the Atomic Scientist* (May 1981) 48 -53.

To an emerging space power:

- Outer space can only be considered as a partially demilitarized zone (only some kinds of weapons are banned).
- To state that a condition of partial demilitarization is synonymous with peaceful uses is an intentional obfuscation of the issues meant only to sow confusion in the definition of what constitutes a peaceful use.
- A peaceful use is clearly a non-military use. It is also clearly a non-weapon use. Problems arise in part because people confuse the existing space regime with peaceful uses of space. The existing space regime, is not a peaceful regime, and it is time this fact is recognized.

IV - Suggested Approach to Realizing Peaceful Use

The question is not whether outer space is currently being used for peaceful purposes. Most countries know that this is not so. The question is rather how do we propose to move from where we are (a situation of non-peaceful use) to where we would like to be (demilitarization of space).

It is eminently reasonable and plausible that the ideal goal of peaceful use can be achieved all at once (which is unlikely) or in a phased way (which is more likely). Issues related to these are addressed in this section.

Figure 5.1 depicts the different categories into which the uses of outer space can be classified. International legal regimes governing specific aspects of these uses (such as ABM systems) are also mentioned. Uses of space that require new international regimes or modifications of existing regimes are highlighted. It is clear from this figure that non-peaceful uses cover two broad categories: a civilian non-peaceful use and a military non-peaceful use.

A typical example of a civilian non-peaceful use is a murder or a rebellion in an international space station. A legal framework for this will surely evolve as and when such stations become operational. The body of existing space law can serve as a basis for this evolution.

The major problems that one has to contend with have to do with military uses, which fall broadly into two categories:

- Support military uses, which include the use of reconnaissance, early warning, navigation, communications and other satellites that help military personnel on the ground during times of war or crisis or during routine operations.
- Defensive military uses (since aggressive uses are banned by the UN Charter), which include development, testing, and deployment of weapons.

CLASSIFICATION SCHEME FOR OUTER SPACE ACTIVITIES

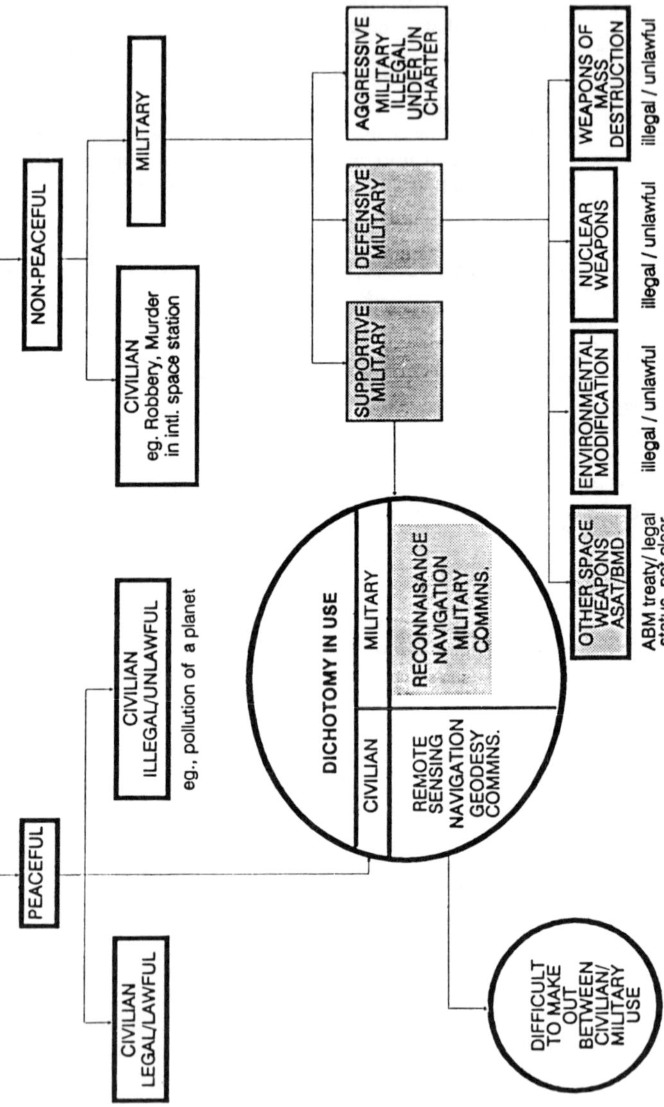

As compared to the first category of support military uses, the conceptual and definition problems related to the second category are relatively simpler. A space weapon can be unambiguously and clearly defined in a generic way. One such definition that has been arrived at and is sufficiently comprehensive to cover most situations that can be visualized is provided in Chapter 1.

However, the first category of supportive military uses of space is more difficult to handle. This is because in practice it is difficult to separate a civilian function from a supportive military function. A remote-sensing satellite can provide information for resources management. It can also provide information for planning an attack. A communication satellite can be used for military communications during a crisis or for civilian purposes. A navigation satellite can be used to target a missile or fix the position of a civilian ship or aircraft. This dual nature of use complicates the clear definition of terms. In addition there are certain other operational factors. Reconnaissance satellites used by the superpowers help in ensuring compliance with the SALT I, SALT II, the ABM and the INF treaties. They thus serve a peacekeeping function. Navigation satellites are closely coupled with SLBMs. They are thus linked with strategic weapons on the ground. Similarly, communications satellites are routinely used by the military and are used for aiding strategic forces on the ground. This linkage of the supportive military function, with global strategic concerns problems has to be taken into account in any approach toward total demilitarization of space.

The emerging space power that objectively analyses the various uses of outer space sees these two aspects for supportive military use and defensive uses of space as two distinct elements in the non-peaceful uses of outer space.

Support military functions that are carried out from space are closely linked to strategic weapons and to the preservation of the strategic balance and have been going on since the very beginning of the space age. The development, testing, and deployment of weapons in space or weapons stationed on the earth directed at space objects are a relatively new element. They add an entirely new dimension to the scope and extent of the arms race between the superpowers. The problems posed by these are more dangerous and require more immediate attention.

To an emerging space power, it is imperative that immediate steps are taken to prevent the introduction of a new generation of weapons into a new arena "outer space." If this step is not immediately taken, "space weapons" will soon become another component of the strategic triad. Negotiations would become more complex, issues more difficult to resolve, and verification problems even more stringent. Rather than trying to address issues related to the entire gamut of space militarization, hard-headed practical, multilateral, or bilateral negotiations on the banning of space weapons are the first priority. Once this has been carried out and the danger of new strategic weapons in a new arena eliminated, issues related to other aspects of militarisation could be taken up.

In the second phase, which follows the elimination of space weapons, complications arise because of several factors:

- Supportive military functions are to some extent coupled with strategic forces; their elimination and reductions is linked to some extent with reductions in strategic arsenals.
- A verification function is necessary, and it is difficult to see how the verification/peacekeeping function can be clearly separated from military support functions.

The way out of this dilemma would be to internationalize the verification function so that a third independent body could serve as an adjudicator. Coupled with this measure, progress in strategic disarmament (if it happens) should be coupled with the regulation of support military activities in space. A non-weapon regime that has hopefully preceded this phase would involve verification. A similar verification provision could be extended to include the support military functions. The number and capability of support military satellites would be fixed as a part of regular arms control agreements. Coupled with and independent verification mechanisms for detection of violations, this would provide the necessary base for building up mutual trust in the system.

V - Approaches for Elimination of Space Weapons and Associated Problems of Definition

To an emerging space power, the issue of the elimination of space weapons is of paramount importance. There are, of course, several route through which this goal can be realized. Each route has associated with it problems of definition of a number of critical terms.

The first approach is one through which all categories of "space weapons" can be eliminated. A typical example of this generic approach is the Antarctic Treaty. It prohibits all military activities in Antarctica. The geographical area covered is also clearly defined in the treaty as the area lying south of 60°S latitude.

A similar generic approach prohibiting the development, testing and stationing of space weapons would be a major step toward realizing a peaceful use of space.

To come to terms with this approach a clear unambiguous definition of "a space weapon" is needed. Such a definition should be comprehensive enough to cover both ASAT weapons as well as BMD weapons, which are the major sources of concern to the international community today. A general definition of a space weapon is provided in Chapter 1.

The characteristics of weapons in general and space weapons in particular are:

- Like all weapons, space weapons should be capable of destroying, damaging, or otherwise interfering with the normal functioning of the object that is the target of attack.
- Unlike other conventional weapons, a space weapon (which could also be based on land, sea, or in air) is directed against an objects in space. This

object could be a satellite, or a ballistic missile flying through space or its nuclear warhead.

Unlike other weapons, a space weapon could be stationed in space or flying a trajectory through space and directed against objects on the earth, in the air, or in space.

Because one of the characteristics of a space weapon is that it is directed against an "object in space," it is necessary to enunciate rather clearly when an "object is in space." One method of doing this, which is rather obvious, is to delineate or define space. To draw a parallel with the Antarctic Treaty, we need to define the equivalent of 60°S latitude for "space."

Many approaches have been thought of to delineate space (see Chapter 2). One approach to this definition imposes a minimum altitude limit (a 100-km altitude is often suggested as the lowest boundary of space). This approach involving a clear spatial delineation has much merit because all foreseeable ASAT and BMD weapons (at least the ones not covered by the existing treaties) would be clearly covered by this definition. Any object that is therefore at an altitude of more than 100 km would qualify as an object in space. Satellites, relevant portions of the boost, post-boost, and midcourse trajectories of ballistic missiles would be covered. To cover pop-up systems, "weapons stationed in space" could be defined to include weapons whose trajectories are in practice above the specified altitude. This altitude could be fixed with respect to any of the standard earth models that are available such as the Goddard earth models.

The advantage with this approach is that it is clear, simple, and unambiguous. Verification problems are to some extent simplified.

However, in spite of extensive debate in the UNCOPUOS, the spatial approach toward the definition of space has eluded consensus. Other definitions including functional ones have also eluded consensus.[13] The major reason is not because the definition of space is difficult. The problem arises largely because of the strategic concerns of the space powers, especially, since ballistic missiles fly a large part of their trajectory through space. These concerns are understandable in UNCOPUOS, which is not supposed to deal with disarmament questions, but a spatial definition of space may not pose any special difficulties to the space powers in a multilateral disarmament forum. This is so, because the subject of discussions are space weapons and the issues addressed are strategic issues. In addition, the status of ballistic missiles flying through space is one of the major items of interests for arms control discussions. Therefore an altitude limit definition for the lower boundary of space, such as a 100 km lower limit, should not be ruled out just because there has been no agreement in UNCOPUOS.

In case the altitude limit definition of space runs into problems, are any alternative approaches possible? One possible approach would be not to define space directly but rather

[13] UNCOPUOS, Background paper, "The Question of the definition and/or delimitation of outer space," A/AC.105/c.2/7/Add. Jan. 1, 1977.

try to define space indirectly by defining "objects in space" -- the subject matter of space weapons. Since objects in space are both targets as well as weapon systems, the definition of an "object in space" should be comprehensive enough to include the weapon systems of immediate interest -- ASAT and BMD weapons. This poses additional definition problems, which, though not difficult to overcome by themselves, make the issue of definition more complex.

ASAT systems are directed against satellites, so "the object in space" is a satellite that is in orbit around the earth. This characteristic, namely "in orbit around the earth," can be used to define one characteristic of an object in space. This would take care of weapon systems directed against satellites and weapon systems in orbit around the earth.

The "orbit characteristic" definition of an "object in space" does not take into account objects that spend only a portion of their trajectories in space. Because this is a characteristic of both ballistic missiles and some weapon systems such as pop-up systems, ways of overcoming this lacuna have to be found.

For this purpose certain characteristics of ballistic missile/pop-up systems could be used for purposes of definition. Some of these identifiable and to a large extent verifiable characteristics are:

- A ballistic missile/pop-up system is in a ballistic flightpath for most of its trajectory. If necessary the term "most of its trajectory" could be defined more quantitatively either in terms of flight times or as a percentage of the total flight time or flight distance or both. The flight time in ballistic trajectory is normally more than about 70 percent of the total flight time.[14]
- A large part of its flight path is in an area where the atmospheric density is low. Even though this may pose special problems due to diurnal, seasonal, and solar activity variations, one could consider putting an upper limit to this density. In other words an object that has a large portion of its flight path in an area where the air density is less than some value can be considered to be an object in space. This mode, however, poses problems of verification. It could be used as an additional parameter if necessary, but not as the sole parameter.
- The burn out velocities or the final velocities achieved by the object could be used to distinguish between objects in space and objects not in space. For ICBMs these burn-out velocities range from about 6.5 km/sec to 7.0 km/sec. Similar values could be arrived at for IRBMs, SLBMs, and other systems if they are considered to be objects of interests. When we talk about velocity, it is necessary to specify both a magnitude and a direction. However, since the direction of the velocity vector determines both the range and the time in ballistic trajectory, even a straightforward velocity definition along with a time of flight parameter may be adequate.

[14] Report of the American Physical Society Study Group, "Science and Technology of Directed Energy Weapons," (April 1987), Fig. 2.1, p. 42.

> The ranges of these missiles could also be considered as an additional parameter. An ICBM would typically have ranges from, say, 3,000 - 10,000 km, IRBMs, 1,000 - 3000 km etc.[15]

An appendix to this Chapter addresses this issue in greater detail.[16] Table 5.1 which has been prepared from the details given in the appendix, gives maximum ranges, optimum injection angles, times of flight, and maximum altitudes for different end-of-boost velocities. This is shown in the form of a graph in Figure 5.2.

Table 5.1

BALLISTIC TRAJECTORY CHARACTERISTICS

Velocity ratio Vf (Vi/Vc)	Angle of injection for maximum range	Maximum range (km)	Time in ballistic flight (seconds)	Maximum altitude (km)
0.1	44.85	63	114	15
0.2	44.41	260	232	64
0.3	43.6	600	360	146
0.4	42.5	1110	500	263
0.5	40.89	1826	662	422
0.6	36.65	2819	861	620
0.7	35.53	4210	1105	860
0.8	30.96	6242	1428	1122
0.9	23.55	9538	1871	1312

Notes:

(1) Vi = injection velocity
Vc = circular velocity of satellite at injection altitude

(2) These calculations are based on impulse velocity being imparted on the surface of the earth.

(3) Air drag effects are neglected in this simplified approach.

(4) The equations used are outlined in the appendix.

---

[15] UNCOPUOS.

[16] S.S. Chin, *Missile Configuration Design* (New York: McGraw Hill, 1961), Chapter 4, Section 4, 9, p. 87; J.W. Cornelise, H.F.R. Schoyer, and K.F. Wakker, *Rocket Propulsion and Spaceflight Dynamics* (Belmont, CA: Fearon, 1979), Chapter 13, Section 13.2.5.

Fig. 5.2

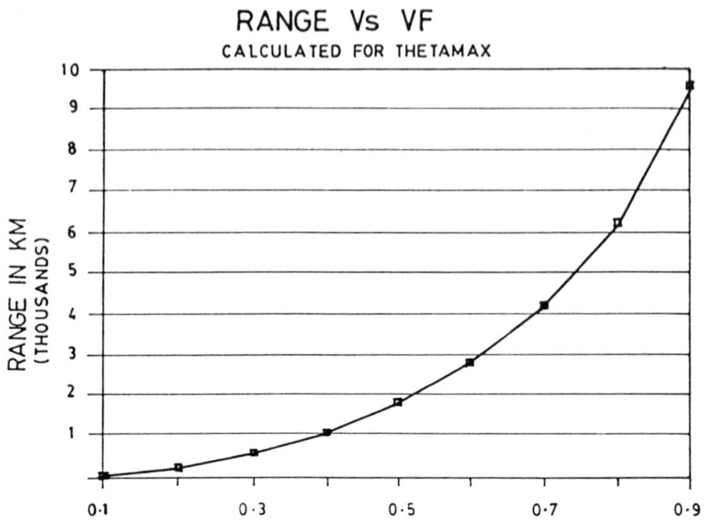
RANGE Vs VF
CALCULATED FOR THETAMAX
RANGE IN KM
(THOUSANDS)
10
9
8
7
6
5
4
3
2
1
0·1
0·3
0·5
0·7
0·9

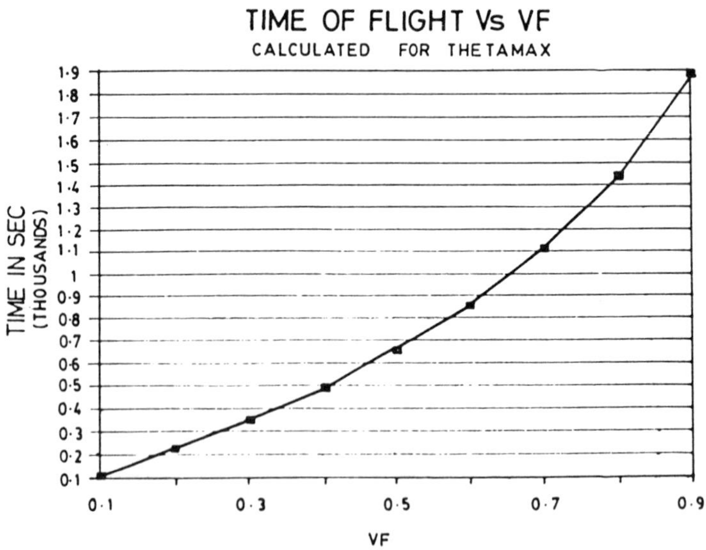
TIME OF FLIGHT Vs VF
CALCULATED FOR THETAMAX
TIME IN SEC
(THOUSANDS)
1·9
1·8
1·7
1·6
1·5
1·4
1·3
1·2
1·1
1
0·9
0·8
0·7
0·6
0·5
0·4
0·3
0·2
0·1
0·1
0·3
0·5
0·7
0·9
VF

Depending on what categories of missiles we want to include as being of interest to space weapons negotiations, Table 5.1 could be used to define missile parameters that are of interest. As an example, if missiles of interest to space weapon systems have a range greater than 2,000 km for a space weapons treaty, then from Figure 5.2 the following parameters of the missile can be used to characterise it as an object in space:

- Its flight time in ballistic trajectory is at least 700 seconds (if time for boost and re-entry is put at about 200 sec its flight time in ballistic trajectory works out to be about 70 percent of its total flight time).
- Its cut-off velocity at end of boost is greater then 4.4 km/sec.

These numbers are illustrative rather than final since factors like air drag are not included in these calculations. More refined and accurate numbers can be arrived at for actual negotiations depending on what missiles fall within the purview of space weapon systems. This, however, does prove that one can arrive at reasonably unambiguous technical parameters that are verifiable to characterize "an object in space."

The time in ballistic trajectory is also extremely relevant because for BMD systems the entire configuration of the defense system is critically dependent on this parameter.

Based on these criteria, an object in space could be considered to have some or all of the following characteristics:

- It is in orbit around the earth.
- It spends a large portion of its flight path (can be defined in terms of actual flight time or as a percentage of the total flight time) in a ballistic trajectory.
- It spends a large portion of its flight in an area where the air density is lower than a pre-specified value.
- It achieve a final velocity of about 6.5 - 7 km/sec. If IRBMs of greater than 2,000 km range are to be included, this velocity limit could be about 4.4 km/sec.

Some or all of these characteristics could be used to determine when an object can be deemed to be in space.

The third approach toward the question of weapons in space would be to consider each kind of weapon system separately. In this approach ASAT weapons would be treated separately and BMD weapons separately. Other categories of space weapons would be dealt with as and when they pose a threat. Definition problems in this mode would be less difficult. The targets of interest could be easily defined for an ASAT treaty, and all ASAT weapons can also be defined using the approach outlined in Chapter 8. Similarly, a ballistic missile can be defined and weapons directed against them banned. This approach of separating ASAT and BMD weapons would, however, pose some special problems. This arises from the close technology coupling between ASAT and BMD weapons. This linkage makes it difficult to verify a selective ban on only one of either ASAT or BMD weapons. This would not be a preferred approach to take unless the international community wants to limit only one kind of weapon system or in some way the ASAT and BMD weapon systems are treated as a package. If the package approach is adopted, it would practically amount to an approach similar to the second approach suggested earlier.

The fourth approach would be to take each weapon type separately and deal with it. Thus kinetic energy weapons, beam weapons, particle beam weapons etc., which serve both ASAT and BMD functions, could be dealt with separately. These approaches would try to use technical characteristics such as beam energies, laser pulse duration, fluence, kinetic energies or velocities, to achieve the required objectives of defining, when their use could be considered a weapon. The targets themselves could also be identified as a satellite or a ballistic missile using functional elements outlined weither in approach one or approach two.

In all approaches it is mandatory that components of weapon systems are also defined. Chapter 8 covers this in great detail.

To an emerging space power the first approach where a clear spatial definition of space is available would be preferred. This approach is generic, and it has the advantage that it completely eliminates the possibility of development of space weapons by all countries. A clear delineation of space also facilitates verification.

The second approach, although not so comprehensive (only ASAT and BMD weapons are covered), would still be acceptable if problems associated with definitions could be resolved. A number of approaches toward evolving these definitions exist. However, this is not as comprehensive an approach as approach one. Since ASAT and BMD weapons are the weapons of immediate concern, this could be a pragmatic approach to take, if the definition approaches suggested either singly or in combination can be used to characterize an object in space.

The third approach of separating ASAT and BMD weapons should be avoided at all costs in view of the close technology coupling the two classes of weapons. Eliminating one without eliminating the other would cause more problems and pose difficult problems of verification.

The fourth approach of treating the weapons by type (such as kinetic energy weapons, beam weapons, particle weapons) would also not be acceptable, because by selective banning of some categories, international security is not enhanced. Emerging space powers may also not adhere to such a treaty because they would deem it discriminatory. In the long term it would create precisely the same concerns that has happened in the area of nuclear weapons and missiles -- proliferation concerns, prohibitions on technology, and discriminatory embargo regimes.

To an emerging space power, space is still a weapon-free arena. A comprehensive ban on weapons would provide a truly non-discriminatory regime for all countries. Selective banning of space weapons or regimes that regulate the number of weapons etc. would result in a discriminatory regime. Whereas such a regime may be successful in the short term, national compulsions could force many emerging space powers into embarking on their own space weapons program. Problems similar to what is happening in the area of nuclear weapons, missiles etc., would repeat itself. Whereas individual short-term self-interests of space powers would be preserved by a selective approach toward elimination of space weapons, the long-term interests of the whole world would be jeopardized since the road to embark on a space weapons program is equally open to all countries. The resultant loser is, of course, world peace and world security.

VI - Technical, Institutional Mechanisms for Ensuring Peaceful Uses

Once basic approaches to the question of the elimination of space weapons have been worked out, a major hurdle toward the eventual demilitarization of space will have been removed. The question of regulation of non-weapon military activity, however, will still remain.

To grapple with problems of verification of a space weapons agreement and of "regulated" support military activities, certain technical and institutional issues have to be addressed. The technical characteristics that could be used for verification -- especially of components -- of weapon systems are outlined in Chapter 8.

In addition to these parameters in both phases of the negotiations outlined earlier, certain technical parameters and some institutional mechanisms could help in separating military and non-military uses.

The technical characteristics of military satellites are fairly easy to define:

- They are normally radiation hardened.
- Weapons would require large amounts of power. Some kind of radiation or physical force would be required for use as a weapon. Most of these parameters that are weapon specific can be fixed quite clearly.[17]
- Telemetry encryption would be a normal feature.
- Radio-silence except over certain areas would be an indicator of a military function.
- Orbit maneuvers and orbit maneuvering capability are usually needed.
- If the data from the satellite is not available freely or its services are not accessible to all countries, the satellite is likely to be a military satellite.

Thus it is fairly easy in a functional sense to separate military and non-military satellites. To facilitate and support such distinctions, there is already a convention, "Convention on the Registration of Objects Launched into Outer Space."[18] This convention requests states launching space objects to provide information on the functions and capabilities of any object launched into outer space. This convention could be strengthened to provide additional

[17] Y. Velikhov, R. Sagdeev, and A. Kokozin, *Weaponry in Space: The Dilemma of Security* (Moscow: MIR Publishers, 1986); S. Chandrashekar, "The Use of Lasers as Anti Satellite Weapons -- An Approach toward Limitation," *Space Power*, 7, no. 3/4 (1988), 311 - 25.

[18] Convention on Registration of Objects Launched into Outer Space, UN General Assembly Resolution 3235 (xxix and annex), adopted on 12 Nov. 1974.

information on power, power sources, amount of fuel, and the nature of the components used in the construction of the space object.

International tracking of the spacecraft, monitoring the telemetry and other forms of electromagnetic radiation from all satellites, could be carried out by suitably identified international bodies. These would to a large extent determine that support military functions are in accordance with regulations in force at that time and that launches of space-based space weapons do not indeed take place. If necessary, inspection of launches could also be conceived of, as a part of this process. This entire exercise could be carried out by an independent body or a neutral group of countries. From this phase, to proceed to total demilitarization would be a much easier task. A UN study on verification using satellites addresses many of these issues in detail.[19]

One important criterion that really helps in making this distinction between military and non-military uses is the free availability of data and space services from the space object of interest. This could be used to distinguish between a military and a non-military function. International cooperation in the use of the spacecraft or satellite would be an additional indicator of peaceful non-military use.

For military uses to be effective, access to data and services from satellites should be restricted. This is the only mode in which the "preferential advantage" necessary for the fulfilment of a military function is maintained. If data and services are available to all, the preferential advantage to any one party would be lost. This precludes preferential use by any one party and facilitates equal use by all parties. If the same reconnaissance data was available to two parties, their approaches toward military confrontation based on politically motivated or imaginary actions of the opponent would be entirely different. In addition, if this information was available from an international body that was neutral, preemptive action by any one party would result in international condemnation with resulting consequences. This would not eliminate all problems between countries completely, but it would be a strong deterrent.

Alternatively, if the service provided by military navigation satellite were to be available freely, all parties would be able to use it equally. In such a situation it is also possible that the technical capabilities of the space segment will be intentionally downgraded so as to preclude possible military use. This in turn would reduce use of space for military purposes and make attacking "objects in space" not very relevant.

One could, of course, argue that "openness" could also result in misuse, especially if a "technically" stronger country uses this capability to attack a weaker country, that does not have the capacity to utilize the service fully.

This is, of course, quite possible. However, situations where this is likely to happen will happen, anyway, even if the space service were not available openly. To ascribe to

[19] "UN Study on the Implications of Establishing an International Satellite Agency," UN document A/AC206/14, 1981.

"openness" and "transparency" in space, responsibility for such misuse is illogical, because this "openness" would not solve the problems related to basic political perceptions and motivations of countries. These require different methods for conflict resolution. By making impartial information and services available, the concept of international openness will provide independent, relatively unbiased factual information. This could reduce tensions, serve as a method of exercising international control over military adventurism, and eventually force countries into the path of negotiations for solving problems. If openness, however, is not coupled with genuine disarmament and is used as a measure only for exercising control over "potential" space powers as is happening in the nuclear and recently in the missile field, then it would just be one more gimmick in a discriminatory space regime. Hopefully this would not happen if the political will exists, for genuine disarmament.

If such a scenario were to emerge and an international regime evolve for regulating "non-peaceful" activities with the eventual goal of total demilitarization of space, emerging space powers would be able to grapple more realistically with the inherent contradictions between their international and domestic goals. If such a development does not take place, more countries will begin to use space for military purposes and eventually more countries would move toward the deployment of space weapons.

VII - Promotion of Peaceful Uses to Reduce Non-Peaceful Uses

In yet another possible scenario, rather than try to focus on "space weapons" and "regulated military uses" of space (phases one and two of the scenario outlined earlier) one could adopt an approach that promotes the peaceful uses of outer space. Such an approach by concentrating on "peaceful uses," would indirectly focus attention on the illegitimate nature of military activities. This method of dealing with the problem has been actively pursued as an alternative to discussing questions related to space weapons in the UNCOPUOS.[20]

The characteristics of peaceful uses that are identifiable through an institutional mechanism such as treaty obligations are:

- Openness of information on the nature and characteristic of the object; this could include technical information also.
- "Transparency" in use, which translates very simply into "everybody can see how an object is being used."
- "International cooperation," which means that all countries desirous of participating in the activity or in using the objects of interest are in a position to do so.

[20] "Revitalising the Committee on the Peaceful Uses of Outer Space," Report of the Committee on the Peaceful Uses of Outer Space, General Assembly official records, supplement No. 20 (A/40/20), New York, 1985, Annex-II, pp. 23, 24.

The UN Registration Convention, which makes it mandatory for states parties to the treaty registering their satellites with the UN, can be strengthened suitably to provide details about the nature and function of the spacecraft as well as about "transparency of use" and "international cooperation." Lack of data or inadequate compliance with these aspects would be an indicator of a military function. The international community would thus get to know about these military uses and focus attention on them.

As some countries have suggested,[21] at some stage this simple starting point could later on be extended to a treaty that guarantees immunity to satellites, installations, and components of these in space, which perform peaceful functions. When a climate of trust has been built up, this could eventually lead to an international regime of "peaceful use."

This approach offers some advantages to an emerging space power. It provides it with an opportunity of keeping its options via-a-vis the military uses of space "open." Whereas this may in theory be beneficial, in practice this would result in a discriminatory regime.

The reasons for this are not difficult to see. The "promotion of peaceful uses in space" and its method of implementation would be very similar to what has happened in other fields such as nuclear energy. Developed countries have not really significantly reduced their nuclear arsenals in spite of the promotion of the peaceful uses of atomic energy and the nuclear non-proliferation treaty. On the contrary, these arsenals have often increased. At the same time this treaty is being used by advanced countries to raise issues related to proliferation among emerging nuclear powers, resulting in a discriminatory regime. One of the dangers in promoting a "purely peaceful approach toward the uses of outer space" is that it could result in a similar discriminatory regime. It may also have very little effect on the activities of the space powers on the military uses of space or the weaponization of space. For "the peaceful approach" to have any meaning to an emerging space power, the promotion of the peaceful uses of outer space has to be closely linked to the elimination of space weapons as a first step and with the phased demilitarization of space as a second step. Any other approach would in its view not contribute in any significant way to the peaceful uses of outer space.

VIII - An Emerging Space Powers' Definition of "Non-Peaceful"

An emerging space power views the problems of definition from a slightly different perspective. Rather than trying to define the word "peaceful," the approach suggested is that one should rather try to define the word "non-peaceful." The definition of "non-peaceful" would facilitate clear identification of the major areas of concern. One could proceed then with concrete action to limit and eventually to eliminate those "non-peaceful uses."

---

[21] France working paper in Conference on Disarmament, "Prevention of an Arms Race in Outer Space; Proposals concerning monitoring and verification and Satellite immunity", CD/937, CD/OS/NP.35, 21 July 1989.

A definition of "non-peaceful" could be along the following lines. A "non-peaceful" use of outer space is any one or all of the following:

- The use of force or the threat of the use of force in outer space or in the earth environment by a space object or being in outer space.
- The use of force or the threat of the use of force against a space object or a being in space by any method or means.
- The use of a space weapon.
- The use of space objects to assist in and aid military operations.

The terms "space weapon," "space," and "object in space" need to be defined if weapons are the first priority. A typical definition of a space weapon is given in Chapter 1. A typical definition of an object in space would involve the following elements. An object in space is:

- An object in orbit around the earth or stationed on celestial bodies or stationed in space in any other manner.
- An object that achieves a cut-off velocity of greater than "X" km/sec prior to entering a ballistic trajectory and that spends more the "Y" seconds in a ballistic trajectory.

This method deals with the definition of space indirectly. Alternatively space can be defined as an area above a certain altitude.

This categorization of "non-peaceful" into supportive military uses and defensive uses permits a phased approach toward the demilitarization of outer space, where the question of space weapons is dealt with first, followed by questions related to the total demilitarization of space.

IX - Conclusions

The absence of a clear international regime governing the peaceful and non-peaceful uses of outer space poses major contradictions between domestic and international policies to an emerging space power. Coupled with lack of progress in achieving genuine "peaceful" uses of space, this will eventually lead to more countries using space for military purposes.

The word "peaceful" has two essential components associated with it. These are an absence of force or conflict and the presence of calm and tranquillity.

The current international regime of space is not a peaceful regime. Activities related to defensive weapons and support military functions can be carried out in outer space. A comparison of the international regime in space with the regime for Antarctica reinforces this interpretation. In addition, international difficulty in agreeing to a common interpretation of the word "peaceful" has been responsible for much of the confusion in this area.

A categorization of the different uses of space along with the legal regimes associated with them reveal two areas of concern: (1) activities related to the weaponization of space, and (2) support military operations in space.

"Weapons in space" are a greater issue of concern because their development and deployment will add a new dimension to the arms race. The regulation of support military activities is connected in some form to strategic arms limitations and global disarmament. The question of peaceful uses should take into account these differences to achieve progress.

The first step is to ban the development, testing, and deployment of space weapons. This could be followed by regulation and control of support military functions as a part of general or strategic disarmament.

To ban space weapons there is a need to define what is a space weapon. The element of this definition should take into account the following:

- It should be capable of destroying damaging or otherwise interfering with the normal functioning of the object, which is the target.
- It can be based anywhere but is directed against a satellite or a ballistic missile flying through space.
- It could be stationed in space or flying a trajectory through space and directed against objects anywhere.

Different approaches are possible toward the elimination of space weapons. In the first approach a generic ban on all space weapons can be arrived at. For this approach to work, it appears logical that to categorize an object in space, some delineation of where space begins is necessary. An altitude limit of around 100 km could be used for this purpose. A second approach, which considers only ASAT and BMD weapons, would involve the definition of "an object in space." For this definition, the functional characteristics that appear most promising to define "objects in space" are:

- The orbit characteristic for a satellite.
- A cut-off velocity criteria for a missile depending on the range of the missiles of interest.
- A time of flight in ballistic trajectory criteria.

The third approach involves dealing with ASAT and BMD weapons separately. Definition requirements for this could draw upon elements of the first two approaches. This is not a preferred approach in view of the close technology coupling between BMD and ASAT. Verification would also be a problem. The fourth approach involves treating each weapon type separately. Thus we would have a beam weapon treaty, a kinetic energy weapon treaty, etc. Here parameters related to damage threshold levels such as kinetic energies, laser pulse duration, laser brightness, fluence, etc., could be used. This approach is not recommended in view of its proliferation potential and selective discrimination.

To an emerging space power, approach one would be the best. Approach two could be acceptable; approaches three and four would be difficult to accept.

An international verification mechanism could facilitate the treaty implementation process and build a climate of confidence. This mechanism could have several components including tracking of space objects, monitoring telemetry, independent observation satellites and if necessary inspection of launchers. A typical scenario of such an international agency using space reconnaissance is outlined in.[22]

In both phase one (elimination of space weapons) and phase two (regulation of military use), a number of technical parameters could be used to help in making the distinction between military and non-military use. These include radiation hardening of the spacecraft, the nature of components used in the spacecraft, the power and power source characteristics, the nature of the telemetry transmission, etc. Free availability of data from the satellite or the free availability of the satellite service to all could be another indicator of peaceful use, because openness, especially through an internationally accepted agency, would contribute to the lessening of tension and prevention of conflicts.

The strengthening of the Registration Convention could also assist in the process of distinguishing between military and non-military use.

As some countries have suggested, yet another possibility is to promote the peaceful uses of outer space using the characteristics of "openness of information," "transparency in use," and "international participation." At some stage this could be translated into a treaty granting immunity to space objects engaged in peaceful uses. This may, however, not prove very fruitful based on international experience in the nuclear field.

A phased approach toward peaceful uses of outer space where space weapons are first banned, followed by progressive reduction in support military activities such as navigation systems seems to be the logical way to achieve progress in this difficult area.

Whereas the world "peaceful" can no doubt be defined to facilitate progress in the peaceful uses of outer space, the suggested approach involves the clear and unambiguous definition of the word "non-peaceful". This definition should cover both a non-military, non-peaceful use as well as a military non-peaceful use.

The major elements of such a definition should include:

- The use of force or the threat of use of force by a space object.
- The use of force or the threat of use of force against a space object.
- The use of a space weapon.
- The use of space object to aid and assist in military operations.

This phased approach would facilitate progress on elimination of "space weapons" and ensure the "peaceful uses of outer space."

---

[22] "UN Study."

Acknowledgments

I thank my colleagues in the Launch Vehicle Programme Office, especially Shri Sreenivasa Setty, for their moral and physical help during the writing of this Chapter. Special thanks to Shri V Sundararamaiah for his help with the programming and the technical and aesthetic contents. Dr. S. K. Sinha's critical comments and his readiness to share his specialized expertise helped significantly in placing the technical aspects presented here on a firm and solid footing. My special thanks to him for this. My secretary K. R. Ambrose has been a tower of strength in patiently making innumerable drafts before bringing this Chapter to its final shape. Without the permission of Prof. U. R. Rao, chairman ISRO, it would not have been possible to have attended the UNIDIR meetings. I am grateful to him for this.

Appendix

Possible Criteria for Characterizing Objects in Space: Problems Posed by Ballistic Missiles

A ballistic missile spends a large portion of its time in a ballistic trajectory -- a trajectory largely governed by the gravity field of the earth. In this it is very similar to a satellite except that in the case of a ballistic missile, its trajectory intersects the surface of the earth at some point. A ballistic missile also spends a large portion of its time at altitudes above 100 km, one of the criteria that is talked about for defining the lower boundary of space.

One of the problems in the definition of a space weapon is to define either absolutely or functionally the terms "object in space," "weapons in space." These definitions are necessary to come to terms with the hard-core elements of discussions on "space weapons."

In case a spatial definition (a definition that puts an altitude limit) of space is not possible, an approach using the functional characteristics of "objects in space" may be necessary. For characterising a satellite, which is an object in space, one can use the characteristic of an orbit. For a ballistic missile, however, this is not possible since it is not in orbit around the earth.

The functional characteristics of a missile that one can use to characterize it as an object in space are:

- The range of the missile.
- The velocity of the missile at its boost cutoff point.
- The time spent by the missile in ballistic trajectory either as an absolute value or as a percentage of the total flight time.

An approach toward how these functional characteristics can be used to characterize when a ballistic missile becomes an object in space is provided in this section.

For a typical ballistic missile profile, see note 16. After the boost phase, which can end at 25 - 80 km altitude and between 60 - 180 seconds after liftoff, depending on the missile, the ballistic missile enters its ballistic trajectory phase. The range and the time of its flight in ballistic trajectory are all determined by the velocity it achieves and the altitude and orientation of the velocity vector at injection. For missiles of concern to space weapons discussions, altitude or height of injection is not a major contributor to either the total flight time in the ballistic phase or the range. At the end of the ballistic phase re-entry occurs at about 80 - 130 km. Neglecting air drag effects during boost phase, the range of a ballistic missile can be shown to be:

(1)

$$R-2r_0\tan^{-1}\left(\frac{(\sin\theta_f\cos\theta_f)}{\frac{1}{(V_f^2)}-\cos^2\theta_f}\right)$$

where

R = range
r_o = radius of earth
θ_f = angle between velocity vector and tangent to the trajectory at point of injection
V_f = V_i/V_c where V_i is the injection velocity and V_c is the satellite velocity at the altitude of injection (normally between 7.5 - 8 km/sec).

Differentiating this expression, the optimum injection angle $\theta_{f\,opt}$, which gives maximum range can be obtained from:

$$V_f^2 = 1 - \tan^2\theta_f$$

Substituting the value of θ_f in (1) we can find the maximum range R for a given V_f.

These results are (i.e., the maximum range) are provided in Table 5.1 for different θ_f values.

The time of flight in ballistic trajectory is given by the following expressions:

For $\theta_f < \theta_{fopt}$:

(2)

$$t_f-\sqrt{\frac{ri^3}{GM}}(2-V_f^2)^{-3/2}[(\alpha-\sin\alpha)-(\beta-\sin\beta)]$$

And for $\theta_f > \theta_{fopt}$:

(3)

$$t_f-\sqrt{\frac{ri^3}{GM}}(2-V_f^2)^{-3/2}[2\pi-(\alpha-\sin\alpha)-(\beta-\sin\beta)]$$

(4)

$$\alpha = 2\sin^{-1}\sqrt{\frac{2-V_f^2}{2}\{1+\sin(\textstyle\sum/2)\}} \quad 0 \le \alpha \le \pi$$

(5)

$$\beta = 2\sin^{-1}\sqrt{\frac{2-V_f^2}{2}\{1-\sin(\textstyle\sum/2)\}} \quad 0 \le \beta \le \pi$$

where r_i is distance of the injection point from the center of the earth, G is universal constant of gravitation, M is mass of the earth, and Σ is the range measured in "angle" units.

The flight times for different values of V_f and θ_f can be calculated from these equations. These are based on the assumption that the velocity in a given direction is provided as an impulse on the surface of the earth. Altitude injection does not change the range or the flight time in ballistic trajectory appreciably.

If total time for boost and re-entry can be assumed to be about 200 sec, we can calculate the percentage of the total flight time that an object is in ballistic trajectory. Whereas this is not suggested as a parameter, since flight time in ballistic trajectory is a more direct indicator, this could also be considered for definition purposes.

As additional information the maximum height that the missile attains for the maximum range condition can also be calculated. One could cross-correlate this with the spatial definition of space. This altitude is derived from the following relationship.

(6)

$$r_a = \frac{r_i}{2-V_f^2}[1+\sqrt{(V_f^2-2)(\cos^2\theta_f)+1}]$$

Where r_a is apogee radius from the center of earth and r_i is injection radius.

The height/altitude reached is given by:

(7)

$$h_a = r_a - r_o$$

where h_a = altitude above surface of earth

Typically these equations and Figures 5.1 and 5.2 can be used in the following way to define objects in space. If the range of missiles covering BMD discussions is assumed to be about 2,000 km, we can characterize any object whose injection velocity is about 4.4

km/sec as an object in space. In addition, the total flight time in ballistic trajectory should be about 670 sec.

If both these characteristics are used, we can cover objects in space comprehensively. If the range of the missile were reduced to 1,000 km, then the cut-off velocity would be 3.2 km/sec and the time in ballistic flight would be 376 sec.

If BMD missiles of interest have ranges greater or lesser than 2,000 km, these values can be changed depending on the missile systems that are to be covered during negotiations.

Generally these broad conclusions are valid. The altitudes achieved with different velocities in the absence of air drag can also be calculated using equations 6 and 7. It can be seen that a 100 km altitude is achieved with a boost velocity of about 1.6 km/sec. Air drag effects when included would lower this altitude.

On a functional basis, if one were to use the altitude criterion a cut-off velocity of 0.3 V_f or about 2 km/sec could be used to delineate the lower boundary of space at approximately 100 km.

The atmosphere through which a missile has to pass complicates the problem in several ways. The ideal values mentioned above are based on the absence of an atmosphere. In actual practice the accelerating missile is subject to drag forces and to aerodynamic heating, setting practical limits to achievable velocities at various altitudes. Some physical understanding of these limits can be obtained based on the following considerations.

For a given thrust of the rocket, acceleration of the rocket will take place till the sum of the drag and gravity forces become equal and opposite to the thrust. In order to estimate the velocity at the end of this acceleration, which will be the maximum velocity at any altitude for a given thrust and for a vehicle whose drag co-efficient C_D and reference area S are known, the following relationship is used:

(8)

$$D \text{ (drag force)} = 0.5\ \rho\ V^2 S\ C_D = F \text{ (Net force)}$$

where
ρ = density of the atmosphere
V = velocity of the vehicle
F = Thrust - Gravity Forces.

This can be rewritten as

(9)

$$D = 0.7\ M^2\ p_a\ S\ C_D = F$$

where
M = Mach number
p_a = outside pressure.

For a typical missile

$S = 6.15\ m^2$
$C_D = 1$
Thrust = 4.5×10^6 Newtons
and p_a can be obtained from standard atmospheric tables.

These results can be used to get an idea of the maximum Mach number that can be achieved on the surface or close to the surface of the earth. This value would be around Mach 3.2 or about 1 km/sec. Cut-off velocities that we are talking about (4 - 5 km/sec) are normally achievable only at altitudes above 20 km.

Thus setting cut-off velocity limits to define space objects also more or less automatically assumes that they have reached a certain altitude. (Thus one of the simplifying assumptions we have made such as impulse velocity rather than graded acquisition of required velocity, is reasonably valid.)

Air drag affects the range, the time of flight, and the height reached. These effects can be computed in an inertial coordinate system assuming that drag and thrust directions are mutually opposite. A parameter that determines the drag effects is the ballistic coefficient, which can be defined as

Ballistic coefficient = $S\ C_D/M$

where S is reference area, C_D is drag coefficient and M is mass of vehicle.

This value for a typical missile will be between 10^{-4} to 10^{-5} m^2/kg. Depending on all these factors (cut-off velocity, angle of injection, shape of the vehicle, altitude of injection), the range, time of flight, and altitude reached can vary.

Based upon these equations and sample calculations, we can prove that the agreements between the ideal calculations and actual realization in a real-life situation are fairly close. This establishes that basic parameters such as burnout velocity and time of flight in ballistic trajectory are valid technical parameters for characterizing objects in space especially insofar as missiles are concerned.

CHAPTER 6: TECHNICAL DEMARCATIONS FOR ASAT AND BMD SYSTEMS

Ashton B. Carter,
Donald L. Hafner, and Thomas H. Johnson[1]

I - Introduction

The problems encountered in distinguishing anti-satellite (ASAT) from ballistic missile defense (BMD) systems lie principally in demarcation, not definition. Perfectly servicable definitions of the two types of weapons can be found in the treaty texts and arms control proposals of the superpowers.[2] However, the common perception is that drawing lines of technical demarcation between ASAT and BMD is quite difficult, because the trajectories of satellites and ballistic missile warheads bear similarities during at least portions of their flights, so that systems appropriate for attacking one may share technical and performance characteristics with systems appropriate for attacking the other. Although there is merit to the common view, the matter is a good deal more complicated, and whether the problems of demarcation are formidable or trivial rests on questions of why such distinctions are being made, what the security requirements are of the parties making the distinctions, and how rigorous the standards of verification must be.[3]

For discussion here, we assume the reason for distinguishing between ASAT and BMD systems is to subject one to more stringent arms control limitations than the other. If neither are to be prohibited, or if both are, it does not matter how they differ from each other (although it might matter how they differ from other things, such as air defenses or non-weapon space activities).[4] Because it would be illogical to limit ASATs if BMD systems are totally unconstrained, the task of demarcation narrows down to cases where BMD systems are limited but ASATs are permitted so long as they lack significant BMD capabilities, and

[1] As this Chapter went to press, Colonel Thomas H. Johnson died after a courageous struggle with cancer. We dedicate this Chapter to his memory -- Ashton B. Carter and Donald L. Hafner.

[2] Article II.1 of the 1972 U.S.-Soviet ABM Treaty defines a BMD system as "a system to counter strategic ballistic missiles or their elements in flight trajectory." A plausible definition of an ASAT would be a system to damage, destroy, permanently disrupt the functioning, or change the flight trajectory of space objects of other states. The Soviet Union proposed a treaty limitation along these lines to the UN General Assembly in August 1983, which in turn presumably reflected a joint draft text worked out by the U.S. and Soviet Union during their (uncompleted) negotiations on ASAT arms control in 1978 - 79.

[3] For additional discussion, see Ashton B. Carter, "The Relationship of ASAT and BMD Systems," in Franklin Long, Donald Hafner, and Jeffrey Boutwell, (eds): *Weapons in Space* (New York, WW Norton, 1986).

[4] For a consideration of distinctions between ASAT and BMD and other systems, see Chapters 7 and 8.

where high-altitude ASATs are prohibited but low-altitude ASAT and BMD systems are permitted so long as they lack significant high-altitude ASAT capabilities.

Precisely what constitutes "significant" ASAT or BMD capabilities must be determined in the context of the security requirements set by the parties to an arms control agreement. Adversaries with large missile arsenals and satellite inventories will in general find it easier to tolerate some overlap in capabilities between ASAT and BMD systems. In contrast, such adversaries may find an overlap less tolerable if they have strategic doctrines and targeting plans that call for highly coordinated missile attacks whose results can be reliably predicted.

Delineating a threshold between significant and non-consequential overlap becomes even more complicated if the arms control regime is multilateral rather than bilateral, with large discrepancies among the missile and satellite inventories of the parties. An ASAT system whose technical overlap made it capable of intercepting 100 missile warheads, for instance, might pose only a nuisance as a BMD against a superpower, not least because a superpower could afford penetration aids to ensure a successful attack against important targets. Yet such an ASAT might be a formidable threat to a state whose deterrence posture rested on an arsenal of no more than 100 warheads.

In setting a threshold between militarily significant and non-consequential ASAT and BMD systems, three criteria are important. The first is to constrain the intrinsic capabilities of the system as deployed. For purposes of discussion here, we assume that a system has militarily significant BMD capabilities if it can intercept more than 10 percent of the ballistic missile force of either superpower, whether those forces are at current levels or reduced by agreement. We select this standard because no arms control regime for ASATs or BMD will be tenable unless it proves acceptable to the United States and the Soviet Union, and a BMD capacity of this magnitude might translate into an ability to defend preferentially a small but important set of military targets against even very large attacks.[5] A standard for high-altitude ASAT significance is less obvious, but we assume a system has military importance if it can destroy 20 - 25 satellites at altitudes of 20 - 36,000 km within a period of 12 hours. This would enable the attack to disable the missile early warning, emergency communications and navigation networks of either superpower, networks important for both conventional and strategic operations.[6] The second criterion is to constrain systems so they do not provide a

[5] The performance standard set for phase one of the U.S. Strategic Defense Initiative, for instance, is to intercept roughly 10 percent of the Soviet ICBM arsenal.

[6] In setting these standards, we are adopting an offense conservative view in evaluating BMD capabilities and a defense conservative view for ASATs. That is, for ASATs, we assume they perform at the level estimated by the side whose satellites are the targets; for BMD, we assume the system performs at the level estimated by the side whose missiles must confront the defense. This amounts to a sure-safe standard for the effectiveness of arms control restraints and generally will favour stricter limitations. If an arms control accord constrains weapons systems so that they appear benign even to opposing states making conservative calculations, it almost certainly will avert deployment of capabilities the agreement intends to prevent. On the altitudes and functions of U.S. and Soviet satellite systems, see Ashton B. Carter, "Satellites and Anti-Satellites: The Limits of the Possible," *International Security*, 10, no. 4 (Spring 1986), pp. 46 - 98, and Paul B. Stares, *Space and National*

foundation from which to "break out" and rapidly deploy a militarily threatening ASAT or BMD capability. The third criterion is to constrain "relabelling," for instance, to prevent deployment of an ostensibly permitted ASAT system that was nevertheless actually capable of intercepting 10 percent of a superpower's warheads and could be "relabelled" and used as a BMD system at the moment of conflict.

We have not mentioned verification, and, in principle, the bases for demarcating ASAT and BMD systems could be selected independently of verification concerns. However, we do not feel there is much point in highlighting distinctions that cannot be readily observed and verified in an arms control regime. As explained below, in our judgment a number of candidates for technical demarcation, such as the size and character of optical sensors, fall into the category of distinctions that are real but non-verifiable for practical purposes.

Conceivable ASAT and BMD systems come in great variety, from those whose components are fully developed and well understood, such as radars and rockets carrying nuclear warheads, to those whose components are still in the transition stage from science to engineering, such as high-powered lasers. Our discussion here begins with technologies that are closest to hand, starting with ground-based BMD weapons. And we focus on demarcations of importance to the primary arms control regime governing these weapons, the 1972 ABM Treaty, with its limitations on the characteristics and deployment modes of regulated BMD systems and its prohibition on giving BMD capabilities to non-BMD weapons such as ASATs. Other regimes are possible, including multilateral arrangements, but focusing on the 1972 ABM Treaty highlights general problems in the two demarcation tasks of concern: staking the boundary between BMD and ASAT, and between treaty-permitted BMD and upgraded, treaty-prohibited BMD.

II - Ground-Based Systems

Ground-Based BMD Systems

A ballistic missile defense engagement is commonly divided into five stages or functions: search and acquisition, tracking, discrimination, interceptor fly-out, and target destruction ("kill"). Search and acquisition generally requires sensors that can scan large volumes of space over very long ranges and reliably detect all potential missile reentry vehicles (RVs). Incoming objects must then be tracked by the original sensor or data on their positions must be handed off to a different sensor for tracking. Tracking for several seconds makes it possible to predict the objects' trajectories and assists in discriminating RVs from other objects (such as remnant booster pieces). Using tracking and other information, the BMD system must then determine which of the objects in the sky are threats that must be engaged. The interceptor must take a finite amount of time to fly to its intercept point in the RV's trajectory, and it must get close enough to ensure that its own warhead destroys the reentry vehicle.

Security (Washington, DC, Brookings Institution, 1987).

New technologies applied to each of these five functions might fundamentally change the capabilities of BMD systems.[7] Search and acquisition are currently performed by high-power, broad-beamed, low-frequency, phased-array radars, which are few in number and vulnerable to attack. Research is underway to replace these radars with various kinds of new subsystems, most of them operating on "other physical principles", primarily infrared or optical wavelength detection. Such sensors can be much smaller and can be mounted on airplanes or on satellites (although the ABM Treaty bans air- and space-based BMD components). Verification of their presence and operation can be made difficult because of their size and because, unlike radar, they are passive sensors: they do not radiate energy but operate on natural or reflected radiation from the target object itself.

The tracking function might also be performed or augmented with passive sensors or with new kinds of optical radars, although more conventional, microwave-frequency radars remain the best option for this function.[8] Efforts to build radars of still higher frequencies (about 30 gigahertz) are hampered both by technological difficulties in generating the beams and by the fact that such frequencies propagate poorly through the atmosphere, particularly through rain.

Reentry vehicles can be discriminated from some kinds of decoys as the attacking warheads reenter the atmosphere at altitudes of 70 - 100 km and atmospheric drag slows the lighter decoys more than the RVs. Low-altitude terminal defense systems take advantage of this natural aid to discrimination, but low-altitude engagement reduces the time available for intercept, which in turn restricts the area on the ground (the "footprint") that can be protected by a particular battery of interceptors.

Decoy discrimination remains the key technical problem of BMD, little closer to a militarily workable solution than it was 30 years ago. Passive infrared and optical techniques remain under intense study but at present seem unlikely to provide discrimination against an opponent with carefully designed decoys. The excellent phase stability of the higher-frequency radars is being studied for use in discrimination also,[9] but such subtle discrimination techniques are especially susceptible to offensive countermeasures. Other possibilities, both active (for instance, reflecting laser light off the target) and interactive (for

---

[7] For a more extended discussion of these matters, see Thomas H. Johnson, "Ground-Based ABM Systems," in Antonia Chayes and Paul Doty, (eds): *Defending Deterrence* (New York, Pergamon-Brassey, 1989), pp. 111 - 131.

[8] The radars currently being designed for the U.S. SDI system's Terminal Imaging Radar (TIR), for example, operate around 10 gigahertz, with about 10,000 transmit/receive modules. They will reach power-aperture products of 1 - 2 million watt-meters squared, insufficient for search and acquisition but excellent for tracking. Search and acquisition require radar power broadcast over a large volume of space, so this capability scales with the product of radar power and antenna area. Tracking requires only a power level appropriate to reach a given range and to overcome noise sources that may mask the target; the antenna area is no longer a major consideration in scaling the performance.

[9] Phase stability refers to the ability to control precisely the pattern of crests and troughs in a radar signal. This property opens the possibility of using the radar signal to create the equivalent of photographic images of the reentry vehicles and decoys.

example, pushing slightly on all targets with directed-energy beams to distinguish light decoys from heavier reentry vehicles) are also being explored.

If directed-energy weapons can be substituted to perform the function of interceptors, then the delays involved in interceptor fly-out become less a problem, replaced instead by considerations of generating, aiming, and focusing the directed-energy beam. Directed-energy weapons with BMD capabilities are far in the future, however. For long-range intercepts at altitudes above 70 - 100 km, interceptor missiles with moderate acceleration and high burn-out velocity are appropriate; for terminal-phase intercepts below 50 km, interceptors with high acceleration and lower burn-out velocities are necessary.

Small nuclear weapons remain the only warheads with adequate lethality for low-altitude interceptions. For intercepts outside the atmosphere, homing vehicles may permit non-nuclear target destruction, although one good way of achieving a cost-exchange advantage with such homing warheads (lofting several small vehicles on one large interceptor) is forbidden by Article V of the ABM Treaty.

The great variety of ways that these technologies and their derivatives might be combined into BMD systems virtually precludes any unequivocal judgments about the effectiveness and military significance of any particular configuration. Nevertheless, by assessing how certain ways of solving defense engagement problems would influence general BMD performance, we can measure the comparative importance of various advances. These assessments can be linked with judgments about verifiability of restrictions on those advances in order to highlight potential thresholds of demarcation.

For illustration, we can consider the Soviet BMD system currently deployed around Moscow, at present the only deployed BMD system in existence.[10] Limited by the ABM Treaty to no more than 100 interceptors, the Moscow system is not militarily significant according to the standard we have set, because it would not interfere substantively with U.S. ability to destroy targets in the Moscow area, certainly not enough to protect the city against a large attack. Any attacker with a moderate-size missile arsenal could simply spend 100 extra warheads to exhaust the interceptors in the system by attacking them (since they cannot be mobile or deceptively based, under the terms of the ABM Treaty), then proceed with the attack against undefended targets. In practice, an attack could probably be effective with fewer than 100 warheads lost to the defense, even assuming a preferential defense, because the current Moscow system is susceptible even to relative crude penetration aids.[11]

---

[10] The United States completed a ground-based ABM site in the mid-1970s but promptly dismantled it. More recently, debate has arisen in the United States over the wisdom of deploying a new BMD system, based on recent technological advances, that would protect against small-scale attacks.

[11] A preferential defense is one in which the interceptors are assigned to defend only a selected number of targets, where the attacker does not know which are well-defended and which are undefended. Penetration aids could include decoys, chaff, and modes of attack intended to confuse the defense system and take advantage of its performance limitations.

In its current configuration, the Moscow BMD is also easily verified by the United States, which can monitor tests of its components and the specifics of its deployment. The key elements of the deployed system, its powerful acquisition radars, are easy to locate. National technical means (NTM) can verify the power-aperture product limit as defined in the treaty (3 x 10^6 watt-meters squared) only within about a factor of three to four for a given radar, but this degree of confidence is sufficient. Any effort to develop mobile engagement radars of such high power that they could themselves also perform the target acquisition function should be detectable at the testing stage.[12] Even if the Soviet Union were to upgrade the system with new treaty-consistent technologies, such as new missiles or high-performance, high-frequency radars, it would not enhance the system's military significance. As long as the number of interceptors remained at the treaty ceiling of 100, they could be exhausted; and as long as the radars were not made mobile or proliferated or hidden, they could be targeted.

A more complicated question is posed by the prospect of BMD breakout -- a sudden widespread deployment (either nationwide or at many key military targets) of technology and components otherwise permitted under treaty limits. Again, the current Moscow system provides an illustration. The system is undergoing modernization, so interceptor and radar production lines are open. Yet without intrusive inspection, it would be impossible to know how many system components of any type may have been produced and stored. Even if they have not been, with active production lines, the Soviet Union in principle could begin now to produce and field a widespread system. And because the United States has no BMD system of any kind fully engineered, let alone in actual production, it would be years before it could take comparable action. Nevertheless, Soviet site preparation and deployment would be conspicuous and time-consuming enough to allow the United States to respond in alternative ways. Effective penetration aids could be deployed on U.S. missiles, for instance, and could provide high confidence of successful attacks in most cases.

In sum, wherever a BMD system is deployed or undergoing modernization, a breakout threat exists at least in some degree. But breakout employing current BMD technologies, in a situation where the missile warhead arsenals of major nuclear states are quite large, is not a potent military threat in the near term and can be made even less threatening later on.[13]

The real issues in upgrades to existing BMD systems have to do with the integration of more advanced kinds of sensor systems, both active and passive.

---

[12] The possibility would remain that a series of permitted radars, such as air defense radars, could be netted together to perform BMD target acquisition, thus evading BMD limitations. Radar internetting would require substantial computational capability, extensive communications, and extensive testing, however. Such testing would involve many radars radiating at once and communicating with one another (and, in the case of optical acquisition, with airplanes or satellites), and would be so extensive as to be detectable and thus verifiable. It would be the electronic equivalent of a broad field exercise.

[13] This conclusion may not be true in the event of extremely deep cuts, 90 percent or more, in offensive forces. Obviously, in such a situation defenses become much more important, and the offensive and defensive forces must be considered together in fashioning an agreement.

Passive optics refers primarily to infrared optical systems that would be airborne, space-based, or launched on rockets at the time of attack. These sensors would either assist or replace radars in target acquisition, tracking, or discrimination. To the extent that they could perform acquisition, such systems could alleviate the defense's reliance on large perimeter acquisition radars, a key vulnerability. To the extent that they could track, such systems could evade any numerical limits on deployed battle management radars and could improve the performance of the entire defense. And to the extent that they could discriminate incoming reentry vehicles from decoys, such systems might permit the terminal defense to engage the attackers at higher altitudes, increasing the effectiveness of the defense by enlarging both the battle space and the interception opportunities.

All this makes passive optics sound quite threatening. Nevertheless, our judgment is that such components are not good prospects for control in a BMD treaty regime, for three reasons. The first reason has to do with verification. The use of passive optics in partial or even full-scale testing could be masked by extensive, simultaneous radar testing over a wide area, especially if the test telemetry were encrypted. An airplane containing the optics (and its encrypted messages) could be explained as part of an alternative command system. One might never know that optical signals were being used for acquisition or battle management, and one certainly could not tell how much the system logic depended upon optics rather than radar. Trying to infer whether passive optical components were being used by simply monitoring the test targets engaged by the defense would necessarily involve very deep guesswork, and the presence of optical sensors on board test aircraft could not be verified without placing observers on the aircraft.

The second reason stems from provisions of the ABM Treaty. If optical systems are used only as adjuncts to improve the performance of radars (for instance, by eliminating problems of radar blackout that results from ionization produced by nuclear detonations), then it is not clear that their use is forbidden by the treaty. In any case, use as an adjunct during tests may be impossible to distinguish from use as a substitute for radars -- indeed, use of optical systems may be impossible to detect at all, as noted above. Hence, the situation might not be improved even if the ABM Treaty were replaced by a more stringent agreement.

The third reason has to do with military effectiveness. Optical systems that assist or even replace radars would not make a BMD system significantly more difficult to overcome, as long as a ceiling remained on total interceptors. Nor would performance improvement in either acquisition or battle management change the basic calculus of BMD breakout; a combination of enhanced penetration aids and larger offensive forces could maintain confidence in deterrence. In contrast, a passive optical system that could actually discriminate decoys from reentry vehicles well outside the atmosphere would indeed be worrisome, because it would cast doubt on some penetration techniques. But experimentation on passive infrared discrimination has been underway for some time now, without clear prospects of success against attacks tailored to counter such optics.

Nor do the prospects seem good for major improvement through the use of active optical components as sensors. These would consist primarily of lasers of various wavelengths, from infrared through visible to ultraviolet. Such lasers could be airborne, space-based, rocketborne, or even, for some battle management scenarios, ground-based.

Because of their inherently narrow fields of view, lasers are of no use in target search and acquisition. Nor do they offer any strong advantages over radars or passive optics as battle management sensors, since the benefits of coherent wavelength and narrow beam width are somewhat offset by the additional command and control problems of pointing and tracking the beam and by the problems lasers have in coping with atmospheric contamination and adverse weather.

But lasers cannot yet be excluded for the third application of new optical components, active decoy discrimination. Because of problems with weather, it is unlikely that such lasers would be ground-based, but it remains conceivable -- though far from demonstrated -- that airborne, space-based, or rocketborne lasers could be used to identify some large fraction of decoys in an attack, by illuminating all incoming objects and analyzing the results. If reentry vehicles can be unambiguously discriminated at long ranges, it might be possible to destroy them outside the atmosphere (by hit-to-kill or nuclear-armed interceptors) at exchange ratios favourable to the defense. Augmented by such advanced active optical systems, a BMD system based essentially on existing components and technologies could become much more threatening, and the possibility of breakout would become a more disturbing prospect.

Arguably, active optical discrimination technologies would confer a whole new BMD capability and thus constitute new BMD components, ones that are banned under the ABM Treaty if they are air- or space-based, and subject to negotiation prior to deployment if they are ground-based and rely on "other physical principles." Moreover, they are likely to be more readily verified than passive optics. Developing a reliable discrimination capability with such lasers would require extensive tests against moderately large clouds of decoys. Because ICBM tests with multiple decoy releases are rare, close monitoring of missile tests, particularly using space-based sensors, should be able to detect the scattered radiation from the discrimination lasers (even though they are of low power compared with laser weapons). Unfortunately, because a successful test series would itself quickly provide a base for deployment, ease of verification may be offset by the ease of breakout with such technologies.

The one candidate being seriously discussed as a replacement for interceptor rockets in an advanced-technology, ground-based BMD role is directed-energy weapons. Within this broad category, only high-power lasers require serious consideration. Ground-based charged-particle beams need not be considered because they cannot propagate far enough, and ground-based radio-frequency weapons operating at power levels we might realistically expect may not be capable of inflicting adequate damage on incoming reentry vehicles. However, even high-power lasers are not an imminent threat. As the 1987 study of the American Physical Society explained, laser weapon development will require many years to demonstrate an BMD capability, many more to create a base for deployment.[14] Meanwhile, BMD-related limitations on the characteristics of ground-based lasers seem premature; in any case, such limitations would not be verifiable without site visits, though certain characteristics of tests conducted with the laser might be verified.

---

[14] N. Bloembergen et al., "Science and Technology of Directed Energy Weapons," *Reviews of Modern Physics*, No. 3, Part 2 (July 1987).

Ground-Based ASAT Systems

We now consider ground-based ASATs and the problems of demarcating them from BMD systems.

Not all ASATs have BMD potential. The coorbital device tested by the Soviet Union from 1968 to 1981, for instance, was too massive and sluggish to have any missile defense capacity, and the U.S. air-launched ASAT, although sharing ancestry with some BMD technologies, nevertheless lacked sufficient targeting flexibility to serve as a BMD weapon.

Flexible, effective, ground-launched direct-ascent ASATs, in contrast, would have some residual BMD capability, because the requirements for engaging satellites and reentry vehicles are similar: both types of interceptors must have high burn-out velocities and divert capabilities.[15] Furthermore, verifying that tests of such ASATs do not confer BMD capacity is inevitably problematic, even with relatively full knowledge of the test conditions. Some characteristics of the anti-satellite test engagement will always be somewhat different from a true ICBM intercept, but the verifier and tester may (and probably will) disagree as to whether the test yielded a BMD-relevant capability. In circumstances where BMD systems are tightly constrained or banned, ASAT tests could be a potential breeding ground of difficulties; where BMD deployments are permitted but limited (as under the ABM Treaty), ASAT deployments that might provide a base for BMD breakout or "relabelling" are of greatest importance.

Several limitations on direct-ascent ASATs could help sustain lines of demarcation with BMD systems. First, tests of such ASATs could be limited to single-warhead ASAT missiles against single orbital targets. Second, such testing should occur only on identified ASAT test ranges, to simplify verification of the first point. With carefully monitored tests limited to single targets, even highly capable direct-ascent ASAT rockets do not offer more of a base for BMD expansion than do existing BMD interceptors. Of course, if the ASAT interceptors were sufficiently close in performance to BMD missiles, one might argue that ASAT deployments in large numbers were simply a ploy to exceed numerical limits on BMD interceptors and sites. One might then propose, as some have, to limit the ASAT rockets' burn-out velocity or divert capability to prevent such relabelling.

In our view, such limitations on ASAT interceptor performance are unnecessary, unless BMD systems are banned altogether. In the first place, a party determined to acquire a BMD breakout potential could do so more directly by covertly stockpiling BMD interceptors themselves. Using ASATs as a mask to accomplish this purpose seems unnecessarily complex and expensive. And limiting the number of deployed ASAT missiles leaves open the same questions about breakout that limiting the number of deployed BMD missiles has left open. In the second place, the problem of BMD capability acquired through dual-capable ASAT interceptors is much more sensibly and straightforwardly addressed by

---

[15] Divert capability is a rocket's ability to alter its established trajectory in order to intercept a target. An interceptor's trajectory will generally carry it close to its target but will leave some finite miss distance for which this lateral "divert" motion must correct.

limiting ASAT sites, not missiles. Thus one might confine ASAT deployments to only a few (say, three) particular sites and on those sites permit only a few (say, 10) individual launchers. If appropriate, the launchers could be governed by the same restraints applied to BMD systems, for instance, no rapid reload or mobile ASAT launchers.

Matters become more complicated if ground-based lasers are substituted for ASAT interceptors, because lasers appropriate for attacking satellites in low earth orbit (LEO) provide an intermediate development objective en route to both BMD lasers and high-altitude ASATs. The BMD and high-ASAT missions have approximately the same power requirements -- that is, somewhat beyond 100 megawatts -- and consequently, at this stage in their development, present essentially equivalent technical goals.[16] In contrast, LEO ASATs require about 10 times less power. Thus, ground-based LEO ASAT lasers suggest a base for development, rather than for deployment, of a laser BMD system. Because the characteristics and even existence of such lasers are very difficult, if not impossible, to verify confidently, it may be possible to pass surreptitiously into the development stage for the high-power device, in a LEO ASAT guise. One should not exaggerate the danger here, however, since it would take several years of development to achieve the 10-fold increase in power from a LEO ASAT to an entry-level BMD laser, and several more years to accomplish system integration. Construction and testing on the requisite scale would be virtually impossible to conceal. Nevertheless, the issue continues to receive considerable attention and so is worth examining.

A continuous-wave laser, operating in the infrared anywhere from roughly one to 10 microns wavelength (depending upon the choice of lasing medium) would require a minimum power between two and 10 megawatts off the final telescope mirror (beam director) to provide a useful military capability as a LEO ASAT. We define that capability as lethal fluence delivered to satellites at 1,000 km altitude and 1,000 km cross-range from the laser's location, assuming 100 seconds of illumination. A repetitively pulsed ultraviolet excimer laser would require essentially the same power, and a single-pulse excimer (which causes damage by impulse loading) would require an output energy in the range five to 10 megajoules.

For most of the candidate lasers for this mission, devices of this size could be constructed in buildings the size of airplane hangars and powered (in most locations) from existing electric grids.[17] Thus their construction and laboratory testing could be carried out entirely surreptitiously. Operation as an ASAT requires (for all but the very longest, least attractive wavelengths) some form of point-ahead compensation.[18] This compensation is

---

[16] The required laser brightness to destroy a satellite at geosynchronous altitude (36,000 km) with a ground-based laser is approximately the same as that required to kill an unhardened ICBM or SLBM booster at 1,000 km.

[17] Free-electron lasers (FELs) are an exception here, since their accelerators, wigglers, and optics sections would be hundreds of meters long. Although the United States is attempting to build such a laser, there is no evidence of a comparable Soviet program.

[18] A laser beam passing through the atmosphere becomes distorted by air turbulence and other effects, and this distortion must be compensated for if the beam's full energy is to be deposited on the target. A complication arises because in the time it takes the laser beam to get from the ground to a

an extremely difficult problem whose solution will require accurate testing against well-instrumented satellite targets. Whether such tests can be performed completely surreptitiously is problematic, but in practical terms the answer is probably yes.

To find such a test (assuming the test telemetry has been encrypted and disguised), one would look for one of two things: laser light scattered by the atmosphere in the beam's upward passage or laser light scattered off the target. Absent knowledge of the laser's ground location, the first approach would be incredibly difficult and expensive, requiring a fleet of satellites constantly examining the sky over the entire area where a laser site was suspected, looking for a signal whose wavelength would be uncertain over almost two orders of magnitude and whose strength would be uncertain over perhaps four orders of magnitude.

The second indicator, laser light scattered off the target, might seem more promising, because the number of potential satellite targets is smaller than the number of potential laser ground sites. Unfortunately, this verification method would still require development of monitoring techniques (such as direct observation from other satellites) to provide continuous surveillance of all satellites -- including prompt monitoring of all newly launched satellites -- for a signal of highly uncertain wavelength and strength (as before).[19] We doubt that the threat merits such an effort.

Ground-based lasers may indeed represent a real threat to LEO satellites, which may or may not be a good reason to discuss an ASAT limitation treaty (complicated by the verification problems described above). But for BMD missions, lasers of ASAT size merely advance the technology toward some still-distant possibility of performing missile defense. In the absence of a negotiated ASAT ban, cooperative rules on laser development could be useful, but it would be misleading to call this an urgent problem.

One useful cooperative rule would restrict ASAT tests to designated ASAT ranges, which might or might not coincide with existing BMD ranges. Because of the verification problem in locating lasers at non-designated sites, such an arrangement would not provide an absolute guarantee against cheating, but it would promote a cooperative arms control position. A second useful rule would restrict the lasers at such sites to 20 megawatts or 20 megajoules, thereby controlling the advance of relevant BMD technology while providing leeway for ASAT development.

Ground-based laser ASATs would be far more effective if used in combination with space-based mirrors to increase their cross- range capabilities (and the variety of angles from which they could attack shielded satellites). Tests of such systems could be restricted to test ranges and to single-bounce scenarios (only one orbiting mirror involved) against orbiting

---

satellite at, say, 500 km, the satellite will have moved more than 10 m, so the laser must "lead" the satellite just as a hunter leads a duck in flight. Hence, the air whose distorting effects must be taken into account is the air between the laser and the satellite's future position. Measuring the turbulence and other distorting effects in this column of air at the moment of ASAT attack is extremely difficult. Lasers operating at longer wavelengths have less need for such compensation.

[19] The strength of the reflected laser light would be even more uncertain if the side carrying out the ASAT test designed its satellite target to absorb as much of the incident beam as possible.

targets.[20] Such tests, because of their complex geometries, should be easy to monitor when carried out at announced test sites. The space mirrors used in such a system are likely to be so distinctive that tests might prove detectable, even if conducted covertly from secret laser sites. The possibility of detection should act as a deterrent to extensive surreptitious programs.

If ground-based lasers powerful enough for BMD or high-ASAT missions are developed, their size and enormous power requirements will make them virtually impossible to conceal. For instance, a 100-megawatt laser facility would require delivered electrical power of more than one gigawatt, the equivalent of what a moderately large city requires. Since development and testing of a ground-based, high-altitude ASAT laser would simultaneously create a base for deployment of a laser BMD system, if high-ASATs are to be permitted while BMD systems are constrained, restrictions must be placed on ASATs to maintain lines of demarcation. For instance, no space mirror tests at the high power levels should be allowed. No more than one such ASAT site should be permitted (about seven would be necessary for an entry-level BMD capability), and on-site inspection and observer presence should be required during operational testing. A strict numerical limit on such tests (two to four per year) against single target satellites might also be imposed; test quotas cannot prevent the transfer of ASAT test data to surreptitious BMD development, but slowing the pace of ASAT tests can make the transfer more tedious and uncertain.

III - Space-Based Systems

We turn now to consideration of more hypothetical future space weapon technologies and their military applications. A comprehensive analysis of all the schemes that fertile imaginations have spawned would be far too complex to undertake here. Fortunately, reality will be less complicated; only a subset of the technological candidates for space weapons will arrive at serious testing, development, or deployment.

As with ground-based systems, when discussion turns to lines of demarcation that might potentially be drawn between permissible and prohibited ASAT and BMD activities, attention inevitably focuses on the technical parameters of components. This is quite sensible, within bounds, but the bounds must be kept in mind: technical parameters may be pertinent in regulating testing and deployment practices, but by themselves they cannot solve the security problem posed by potential breakout.

[20] Parties to an BMD arms control regime might establish an agreed orbiting test range (AOTR) for tests of BMD systems with space-based elements. For a discussion of this concept, see Ashton B. Carter, "Limitations and Allowances for Space-Based Weapons," in Antonia Chayes and Paul Doty, (eds): *Defending Deterrence* (New York: Pergamon-Brassey, 1989), pp. 148 - 150.

Space-Based BMD and ASAT Systems

One possible mode of space-based BMD would rely on space-based missile interceptors (SBI) for boost- or postboost-phase defense. Such interceptors would have two key subsystems: a homing warhead with on-board homing sensor, guidance system, and divert rocket motors for terminal maneuvers; and axial propulsion to carry the warhead far from the carrier satellite from which it is launched. Test programs to develop these subsystems would have several characteristics. For the homing subsystem, key dimensions of a test program would include: the relative velocity of the warhead and its target at intercept; the extent to which the target duplicated the exhaust plume of a genuine ballistic or postboost vehicle; the degree to which the target's direction of flight and the position of the earth relative to the homing warhead simulated the geometry of an actual BMD engagement. For the axial propulsion subsystem, the key parameter would be the velocity it imparts to the warhead relative to the carrier satellite, called δv ("delta vee").[21]

A capacity for breakout deployment of a BMD system based on SBIs might proceed in either of two ways: development and testing of SBI BMDs, followed by breakout deployment; or deployment of SBI ASATs, followed by rapid upgrade to BMD capability. To address this concern, a number of analysts have suggested that limits be imposed on the δv potential of SBI propulsion subsystems. An SBI must have a δv of at least 6 to 8 km per second to be useful for boost-phase BMD. With a smaller δv, each carrier satellite could make intercepts only in its immediate vicinity, so that a larger number of satellites would be required to provide worldwide boost-phase coverage. The BMD potential of a constellation of SBIs increases both with the number, N, of deployed interceptors and with δv, and is roughly proportional to $N(\delta v)^2$ for many configurations of attacking boosters.[22]

An adequate SBI ASAT deployment, in contrast, need not have a large value of $N(\delta v)^2$: the number of satellite targets for an ASAT would be smaller than the number of missile targets for a BMD, and the time available to attack satellites is much longer than the duration of the boost and postboost phases of a ballistic missile. Thus a rule limiting SBI ASAT deployments to small values of $N(\delta v)^2$ (or some other appropriate function of N and δv representing the constellation's BMD potential) would ensure that they had negligible BMD potential without unduly limiting their usefulness as ASATs.[23]

---

[21] We are assuming for discussion that multiple space-based interceptors are stationed in orbit aboard carrier satellites and launched from the carrier against their targets at the moment of attack. However, the general principles would equally apply to SBIs placed in orbit individually, without a carrier satellite, as envisioned in the SDIO's "brilliant pebbles" scheme.

[22] The radius of action of each SBI is δv multiplied by the time interval available between the moment the ascending booster is acquired as a target and the end of its post-boost phase (when its exhaust plumes can no longer be used for targeting). The square of the radius of action times pi gives the area of the earth's surface "covered" by the defensive SBI. The number of SBIs times the coverage area of each measures the thickness of coverage over each ICBM silo field or SLBM patrol area. Against some configurations of offensive boosters, the constellation's BMD potential is a more complicated function of N and δv.

[23] A limit of a few hundred square kilometers per square second would ensure that SBIs could not intercept more than a few tens of SLBMs and ICBMs.

A stronger constraint on ASAT upgrade would involve placing a separate limit of, say, 1 km per second on δv itself. However, a limit on testing or deploying SBIs at δv's greater than 1 km/sec would not necessarily prevent development of an BMD-capable SBI. Experiments to develop homing warheads and axial propulsion subsystems could be conducted separately, although the subsystems would need to be integrated in a production model prototype. Homing warheads could be tested ostensibly as ASATs against satellites equipped with thrust motors, thus allowing developers to explore the warhead's performance against a realistic target over the full range of BMD closing velocities (that is, the relative velocity between the warhead and the target), even if the SBI warhead were dispatched from its carrier satellite at low δv.[24] A limit on δv alone, therefore, would not guarantee that BMD-capable homing warhead could not be developed. The axial propulsion subsystem could then be developed in ground test-stand firings and in space tests conducted without the homing warhead. In the end, both pieces of a BMD-capable SBI would have been developed.

The case of the δv limit for SBIs illustrates the inadequacy of thinking in terms of capability thresholds for test devices without reference to how the test is conducted. Testing practices and target characteristics, not velocity thresholds, are the most important focus for a treaty regime seeking to place limits on SBI BMD. Thus if it were considered essential to prevent development of BMD-capable SBIs to preserve the buffer against breakout, multiple dimensions of testing practices and targets should be regulated. First, SBI testing could be banned altogether in ASAT or lofted test modes.[25] If SBI ASATs are still to be permitted, then tests against thrusting targets could be prohibited, and an additional limit on the δv of SBI ASATs (1 km/sec or so) might inhibit ASAT upgrading. A restriction on δv alone would interfere with other non-weapon space programs, so it should be paired with a limit on high acceleration.[26] Thus the regime might limit space objects from maneuvering with δv more than 1 km/sec only if their acceleration exceeds 10 times the acceleration of gravity (10 g's).[27]

---

[24] In a space-based ASAT test, a weapon in orbit would intercept a target that is also orbiting rather than ascending from the earth as a missile booster does. By orbiting in opposite directions, the target satellite and the SBI/ASAT carrier satellite would approach each other at high relative velocity. In addition, a satellite target can easily be made to resemble the post-boost vehicle of a strategic missile, which thrusts only intermittently and at low thrust levels. If the satellite target is equipped with a large booster motor, it can also be made to resemble the second or third stages of a strategic missile booster. The SDIO reportedly plans to test a space-based laser called Zenith Star against a thrusting satellite target.

[25] In a lofted test mode, the SBI weapon is not placed in orbit but is lofted into space on a booster for a few minutes while the test is conducted and then reenters the atmosphere. The SDIO plans to test early versions of its space-based interceptor in this manner.

[26] Nonweapon spacecraft regularly blast off from low earth orbit with δv greater than 1 km/sec, but do not require high acceleration. For instance, communications satellites require a total δv of 6 km/sec to move from an initial inclined, low-earth "parking" orbit to their geosynchronous equatorial deployment orbits. Planetary probes depart from low earth orbit with at least 3.5 km/sec velocity change.

[27] Yet another threshold parameter for SBIs has been discussed from time to time: requiring that the homing warhead have a mass no less than some threshold, M. The rationale for such a threshold is that the launch cost of an SBI constellation is proportional to the constellation's total mass, NM.

Another category of space-based ASAT and BMD would substitute lasers and mirror relays for SBIs. For such systems, different technical constraints have been proposed.

The rate at which a space laser weapon can destroy targets at a given range, and thus its BMD capability, is determined in large part by its brightness. Numerically, brightness equals the power of the laser beam divided by the size of the diverging cone into which the weapon's mirror can focus the beam. The power of the laser is its energy output per second, measured in watts. Cone size is measured in steradians, a unit of measurement similar to the square degrees of angle obtained by multiplying the angular width of the cone by its angular height. A threshold on laser brightness is thus expressed in watts per steradian. Brightness is similarly a measure of the lethality of a mirror relay satellite that does not generate its own energy but that collects and refocuses energy generated by a laser on the ground or on another satellite.

If the laser (or relay mirror) emits energy in pulses rather than a continuous beam, its effective brightness is just the energy of each pulse (in joules; one watt equals one joule per second) times the number of pulses per second, divided by the cone size (in steradians). If a pulsed laser could emit extremely energetic but short pulses, a target booster might be destroyed by the so-called impulse-kill mechanism rather than by scorching, as with a continuous beam.[28] Treaty regime rules for such a pulsed laser might therefore define both an effective brightness threshold and a separate threshold on the number of joules per steradian that could be emitted in a single pulse.[29]

Brightness has attracted attention as a threshold parameter because of its intuitive appeal and theoretical elegance, but in fact the BMD capability of a space-based laser (SBL) is more complex than a single parameter can define. The number of boosters of a given type that an SBL can engage in a given period of time depends not only on how long it takes to kill each booster (determined largely by brightness), but on how long it takes to move or "slew" the beam from one target to another (a function of the pointing mechanism's slew angular acceleration, maximum slew angular velocity, and resettle time) and the supply of energy aboard the satellite (chemical reactants or stored electrical source). The acquisition, pointing, and tracking system that directs the beam to the target's location is a complex subsystem whose BMD capability is characterized by its own array of thresholds. Moreover, treaty limits on brightness can be replaced or augmented by limits on laser power and mirror size. These parameters differ in their verifiability and in their importance for constraining

---

Setting a minimum mass threshold would permit tests only of large-mass "clunkers" that would be too expensive to deploy as a full BMD system. One problem with a mass threshold is the ease with which equipment could be tested on "clunkers" and then integrated into smaller SBIs. Another problem is verifiability: the principle of the equivalence of inertial and gravitational mass means that objects of different mass can follow the same trajectories in space, so that mass cannot readily be determined by external observation. The same inconvenient principle lies behind the well-publicized difficulties with verifying the throw-weight of offensive missiles.

[28] In impulse kill, such a large amount of energy is delivered to the skin of the booster in a short time that the skin vaporizes explosively, sending an impulsive shock through the booster.

[29] The interaction of a laser beam with a booster body can be complex, and harmful effects may depend on other features of the laser beam than just the rate of energy deposition per unit area.

development of BMDs and upgrading of other systems. The following discussion uses brightness as an illustrative proxy for a complex set of potential technical limitations on space lasers and relays.

As with SBIs, testing practices for laser weapons are at least as important as the parameters characterizing the test devices. Once again, there are two routes one could take to break out: development of SBL BMD followed by rapid deployment; and deployment of SBL ASATs or SBL air defense weapons followed by upgrade.

BMD defense against an opponent who took measures to harden his boosters against lasers would probably require space lasers brighter than 10^{22} watts per steradian.[30] But lasers much less bright than this could threaten existing, unhardened missiles and thus would cause concern. The amount of heating that currently deployed missiles could tolerate without damage or disruption is unknown (and differs from the amount that the defender could be confident would destroy them). If we assume a space-based laser would be worrisome if it could deposit more than a "sure-safe" flux of 140 watts per square centimeter (1,000 times the flux of the noonday sun at the earth's surface) on a target at a range of 500 km, then an SBL would be BMD-capable if its brightness exceeded 4×10^{17} watts per steradian. Thus the threshold of concern is fully 25,000 times less than the BMD developers' goal of 10^{22} watts per steradian.

To impede development of BMD-capable lasers (if this were deemed necessary), space tests of SBLs with brightnesses approaching 4×10^{17} watts per steradian could be prohibited, including tests in an ASAT mode (at least when such tests involve thrusting targets) and in a lofted mode.[31] To prevent upgrade of SBL ASATs, if they are permitted, the number of deployed SBLs should be limited so that the constellation has negligible BMD capability. The limit might be phrased, for example, as a limit on the number of laser ASATs multiplied by their brightness, analogous to the $N(\delta v)^2$ limit for SBI.

SBLs have also been proposed as a means of discriminating reentry vehicles from light decoys in midcourse flight by bursting, warming, or nudging the decoys. Such an "active discriminator" would have to be quite bright, since it could devote only a short time to each of the many objects it has to illuminate. Though an SBL discriminator might be characterized as an adjunct to a BMD system rather than as a BMD weapon, it would probably be brighter than the threshold for BMD capability of 4×10^{17} watts per steradian, so development of laser discriminators should be subjected to the same constraints as laser weapons.

---

[30] The American Physical Society's Study Group on the Science and Technology of Directed-Energy Weapons suggested that the reasonably rapid thermal structural kill of a hardened booster might require an energy flux in excess of 200 kilowatts per square centimeter. At 2,000-km range, this requires a laser of brightness greater than 10^{22} watts per steradian (see note 14).

[31] Bright lasers and even relay mirrors, unlike SBIs, might be too heavy and unwieldy to test in the lofted mode.

Limits on technical parameters for particle-beam weapons could be approached in much the same way. The lethality of a particle-beam weapon (PBW) is determined by the energy of the particles (in electron volts), the number of particles emitted per second (in amperes), and the size of the cone into which the beam can be focused (determined by the particle accelerator, focusing magnets, and neutralizer). One can define the brightness of a PBW just as for a laser: the power in watts is the particle energy times the number of particles per second (multiplying volts times amperes gives watts), and the cone size is measured in steradians.

The numerical thresholds for BMD capability would be about the same as the laser thresholds if the PBW has to heat its target to destroy it. But particle-beam weapons may affect their targets in other ways, by disrupting internal electronic components. An appropriate threshold on PBWs to constrain this effect would be 1,000 to 10,000 times lower.[32] PBW discriminators would probably need to be almost as bright as PBW thermal-kill BMDs and hence could be constrained within the same threshold.

It should be evident from our discussion of both ground- and space-based BMD weapons thus far that the option of permitting effective BMD systems while prohibiting high-altitude ASATs -- something we noted at the outset was at least conceivable in principle -- cannot be exploited through a set of simple technical parameters. The power requirements for both ground- and space-based lasers in the BMD and high-ASAT missions, as noted above, present essentially equivalent technical goals. Similarly, the δv capacity of space-based BMD interceptors (SBIs) would overlap the ASAT propulsion requirements for reaching geosynchronous orbits. Even though the high-ASAT mission might demand additional capabilities in lasers or SBIs (for instance, greater precision in beam-handling, longer endurance for interceptor flights times lasting hours rather than minutes, targeting systems not dependent on the bright beacon of booster motors, etc.), it may be difficult to ensure that these are not being developed in a regime where open testing and deployment of BMD weapons are permitted, unless more complicated restraints are adopted.[33]

---

[32] Ashton B. Carter, "Directed Energy Missile Defense in Space," *Background Paper of the Congressional Office of Technology Assessment*, OTA-BP-ISC-26 (Washington, DC, U.S. Government Printing Office, April 1984), pp. 29 - 30.

[33] It would help if limitations on space-launch capacities also ensured a comfortable buffer of conspicuous and time-consuming deployment activity between allowed testing of BMDs, or deployment of space weapons for other military purposes (ASAT or air defense), and achievement of a meaningful defense against unhardened offensive missiles. In general, the longer it would take to deploy a defense based on a certain weapon technology, the less need there is to constrain the development and testing of that technology, so the ability to place payloads into orbit would establish a lower limit on the time required to complete a breakout deployment. Unfortunately, this minimum time is quite short, at least for a first-generation defense of limited capability that might be deployed by the United States or Soviet Union. The weight of a breakout deployment would depend on its specific technology and military capability. For example, the SDIO earlier proposed a phase one deployment for the United States that includes a space-based tier consisting of several thousand interceptors, intended to intercept 10 percent of the Soviet ICBM force and involving launch to low earth orbit of 1 - 2 million kg. Follow-on phases involving directed-energy weapons have estimated weights of 7.2 - 18.6 million kg, and far-term deployments as much as 80 million kg. (U.S. Congress,

IV - Verification

The ability of the two superpowers to verify arms control limits on BMD systems and strategic ballistic missiles -- and hence their willingness to enter such agreements -- stems from the steady accumulation of monitoring capabilities and interpretative knowledge over more than 30 years. Monitoring has been helped in some degree by the fact that both sides have conducted their BMD and missile tests year after year in much the same places and manner. Consequently, each side could learn a great deal about the others programs, at bearable expense, by focusing its monitoring systems and data collection efforts on a few test ranges.

Space-based testing of weapons threatens to upset these monitoring practices because orbiting objects do not confine themselves to single test areas. Devising a practical scheme for monitoring arms control compliance in space is therefore not just a matter of identifying advanced remote sensing technologies that could in principle detect laser light or other test phenomena. A treaty regime must also constrain testing practices so that monitoring is operationally feasible and affordable.

A variety of rules could help. An agreement could require, for instance, that all permitted space weapon tests be conducted in traditional ICBM and SLBM test ranges or from an agreed BMD test range. A more restrictive rule would require that all space weapons approaching BMD capability be tested in agreed orbiting test ranges. Other operational aids to verification could include prior notification of certain types of tests, prior description of tests, inspection of test devices on the ground before launch, and restrictions on concealment tactics such as telemetry encryption.

The richest two generic sources of information on spacecraft would probably be telemetry and imagery. Electronic signals emitted by space test vehicles might include beacons for tracking them and radio broadcasts of the output of on-board sensors and

---

Office of Technology Assessment, *SDI: Technology, Survivability and Software*, OTA-ISC-353 (Washington, DC, U.S. Government Printing Office, May 1988), p. 149.) If limitations on space launch rate alone were relied upon to provide a two-year buffer against breakout with a system like SDI phase one, then superpower space launch capacities would have to be under 0.5 million kilograms per year. By contrast, the Soviet Union already places about 0.5 million kg into low earth orbit in about 100 separate launches annually, and the U.S. Defense Department expects Soviet lift capacity to increase steadily over the next two decades, from 0.5 million kg per year to 1 million kg/yr in the early 1990s, and to almost 2 million kg/yr by 2005 (U.S. Department of Defense, *The Soviet Space Challenge*, November 1987). Prior to the Challenger accident, U.S. space launch plans called for placing about 0.2 million kg into orbit annually in about 20 launches, and although the United States has a program for a heavy-lift Advanced Launch System, it could not be operational until the late 1990s. These total launch capacities must be shared among military space programs, civilian missions (scientific, manned, and planetary), and commercial missions (communications, weather, remote sensing). To avoid having to shut down other space programs, lift capacity in excess of their requirements would have to be found for launching a breakout BMD. Nevertheless, superpower launch capabilities are currently large enough that they may not provide significant constraints on BMD breakout potential.

diagnostic equipment. Images could be collected from the ground in sunlight or by illuminating the satellites with laser light or radar; or one satellite could be photographed from another.

The velocity and acceleration of SBIs might be determined from the arrival times and Doppler shifts of their radio emissions, much as navigators use navigation satellite signals to measure their position and velocity on the ground. Ground-based radars could also track test objects if the tests were confined to regions of space visible to monitoring radars. Last, space-track sensors akin to the SDI's Space Surveillance and Tracking System (SSTS) could aid in tracking SBIs.

Because the beams from directed-energy weapons of potential BMD capability would be narrow, it is unlikely that monitoring satellites would be in the beam at the time of a test. Instead, energy reflected or scattered outside of the narrow beam cone (by the edges of the laser-focusing mirror or particle-beam neutralizer, by space junk, by the target, or by the earth's atmosphere) would have to be used to characterize the weapon (laser wavelength or particle energy) and perhaps make a rough guess at its power. Further information about power might come from observing the combustion products from chemical power sources as they were vented into space or from detecting the electrical emanations of electrically pumped lasers, particle-beam accelerators, or rail guns. A laser's wavelength can be determined by collecting scattered light; inspection on the ground or imagery in space could reveal the size of its mirror. This information would make it possible to estimate the size of the cone into which the laser can focus its light; if its power can also be estimated, its brightness can be calculated.

It is not possible to render in advance a general judgment on the quality or adequacy of the technical information that will be collectible from space weapons programs. Plausible treaty regime rules crafted to ease monitoring would constrain the time, place, and manner of tests; ban methods of concealment and deception; and perhaps permit pre-launch inspection of certain payloads. After that, the efficacy of verification will depend on the willingness of parties to an agreement to make investments in technical collection systems commensurate with their concern about the integrity of the treaty regime and the prospects of violations.

V - Conclusions

Drawing lines of distinction between BMD and non-BMD systems and activities is not simple. The two superpowers have been engaged in what has proved to be a ceaseless effort to find and maintain such distinctions since 1968 -- in the SALT negotiations that produced the 1972 ABM Treaty, in the Standing Consultative Commission (SCC) established by the treaty, and more recently in the defense and space negotiations in Geneva. The constant labor devoted to this task underscores its importance, but also its perplexities. And some would argue that the task has recently become even more important, as an increasing number of states acquire ballistic missile and space launch capacities, and thus in principle acquire greater potential for embarking on their own ASAT and BMD programs.[34]

The simultaneous proliferation of new technologies applicable to ASAT and BMD systems and of new states with missile capabilities creates great temptation to forge an arms control regime based on broad generic definitions and technical demarcations, on the presumption that the broader the terms of restriction, the better they will cope with the variety of technologies and the differing levels of technical advancement among national ASAT and BMD programs. We hope the analysis in this Chapter assists the arms control enterprise, not least by offering several cautions about an overly broad approach to limitations. First, one should not forget that technologies in themselves are not components and components are not weapon systems. There is a natural tendency when trying to constrain systems to focus on their constituent parts, and too often in the process, the whole is forgotten and an obsession with the parts takes over. As we argue, it may be neither necessary nor prudent to limit the technologies of some components (such as sensors), so long as the acquisition of a total system in which they are only a part can be curbed in other ways. Overly broad limitations may unwisely constrain beneficial applications of new technologies.

Second, the great variety of ways that all relevant technologies and components might be combined into conceivable ASAT and BMD systems, now and in the future, would almost certainly defeat any effort now to negotiate comprehensive limitations "of unlimited duration". There is no need at the moment to address the more speculative technologies, merely because they excite the imagination of weapons builders. There are challenges enough to be faced in coping with new developments that we can discern on the immediate horizon, and if a negotiating framework is established for adapting limitations as new problems emerge, the more remote and speculative technologies can be dealt with when the time is ripe.

Third, it is well to remember that the price of failing to draft limitations carefully will be the unwillingness of some states to conform to a negotiated arms control regime. Those who would press ahead with a control regime, despite dissent from states with significant ASAT or BMD potential, may derive psychic satisfaction from standing on what appears to be the moral high ground. But a treaty that means well is a poor substitute for one that works well.

The contributions toward stabilizing the strategic relationship among nuclear powers, and thereby reducing the nuclear danger, that have been made by both the 1972 ABM Treaty

---

[34] Although it is generally true that acquiring a ballistic missile or space-launch capacity improves a state's technological base for ASAT and BMD programs, the point should not be exaggerated. Even a rudimentary ASAT system requires significant advances in a broad range of capabilities that are not customarily part of a missile or space program, including a space-track network that can locate and identify noncooperative targets and operational precision in launch procedures, guidance, maneuvering, and sensor components that can reliably deliver a lethal ASAT warhead to a point in space with errors no greater than tenths of a second and tens or hundreds of meters. See Donald L. Hafner, "For the Benefit and in the Interests of All: Superpower Space Programmes and the Interests of Third States," in W. Stutzle, B. Jasani, and R. Cowen, (eds): *The ABM Treaty: to Defend or not to Defend?* (London, Oxford University Press, 1987), pp. 197 - 201.

and superpower restraint in ASAT programs have been substantial. The task of extending these achievements is vital, but vital enough to do prudently -- and we have time available to be prudent.

CHAPTER 7: PROBLEMS OF DISTINGUISHING ATBM/AIR DEFENSE SYSTEMS FROM SPACE WEAPONS

Hubertus M. Feigl

Ballistic missile defense weapons of the ATBM category are intended to provide a defense against medium-range and shorter range ballistic missiles. From a technological point of view, ATBMs can be understood either as a subcategory of strategic BMD systems or as an "offspring" from upgraded air defense systems. Problems of distinguishing between ATBM/air defense systems and space weapons originate from technological and functional overlaps that do not only exist between air defense and BMD systems but also between BMD and ASAT systems. In many cases these overlaps appear as residual capabilities that may meet the criteria of a space weapon. To identify residual capabilities of that kind, the ASAT capability instead of the BMD capability of a particular weapon should be used as a reference. Current technology provides ground-based ATBM and air defense missiles only with a marginal capability to intercept low-orbit satellites. This observation may change with more sophisticated terminal defense missiles becoming available. Whether ATBM requirements may act as development drivers of "exotic" weapon technologies cannot be answered at this stage. But, with the potential advent of X-ray laser weapons and high-energy optical laser weapons, the intercept ranges pertinent to ballistic and orbital targets traversing outer space may significantly increase. Considering the attractiveness of optical laser weapons for meeting tactical requirements, it becomes plausible that air defense attains the role of a development driver toward space weapon applications.

I - Introduction

Outer space has increasingly become an attractive medium for various sorts of military exploitation including the use of weapons.[1] Though the full range of relevant activities obviously has been a domain of the superpowers, other nations have taken steps to join them sooner or later.

Many military uses of outer space are related to the operation of military support satellites that are not destructive in themselves. This led to the development of a variety of devices of destructive operations in outer space. A wide range of weapon systems capable of engaging "traditional" space targets, such as satellites in earth orbits and ballistic missiles

[1] Bhupendra Jasani (ed), *Space Weapons -- The Arms Control Dilemma* (London: Taylor & Francis, 1984).

leaving the atmosphere on their way to target[2] are being investigated (see Table 1.1 and Chapter 6).

In order to clarify the issue to be considered, a general working definition of space weapons is given in Chapter 1. The definition of a space weapon is linked with the definition of the boundary between the air space and outer space. For a full discussion of this, see Chapters 2 and 5.

> *A space weapon is a device stationed in outer space (including the moon and other celestial bodies) or in the earth environment designed to destroy, damage or otherwise interfere with the normal functioning of an object or being in outer space, or a device stationed in outer space designed to destroy, damage or otherwise interfere with a normal functioning of an object or being in the earth environment. Any other device with the inherent capability to be used as defined above will be considered as a space weapons.*

The definition of space weapons focuses on objects in outer space being either the target or the carriers of space weapon. Important examples are "weaponized" satellites and weapons capable of intercepting ballistic or orbiting targets. These fall into two categories: ASAT weapons and BMD weapons, which are specifically designed to intercept targets in space. Whereas the operational requirements of ASATs differ greatly from those of BMDs, both approaches refer to a similar technology basis (e.g., guided weapon technology). Much the same is true with regard to advanced air defense weapons (AD weapons) and "anti-tactical" ballistic missiles (ATBM weapons). The defense requirements to be met by ATBMs are more of a regional character. This may be quite in contrast to the strategic variety of BMD weapons. The technology base that primarily incorporates "traditional" missile concepts of ATBM and air defenses may be supplemented, in the long run, by "exotic" weapon technologies, such as those used in KEW or DEW. Weapons of that kind may attain a considerable importance, notwithstanding that the outcome of these developments is dubious. Considering the overlap of some of the techniques, it is not simple to know the extent to which ATBMs and air defense weapons may be classified as space weapons. In dealing with these problems it should be stressed that the employment option of a particular weapon system has to be carefully weighed against its residual capabilities, which may be those of a space weapon.

II - Identification and Allocation of ATBM Missions

The quest for effective defense against short-range missiles (but not exclusively) corresponds to regional defense requirements in Central Europe, which are extremely diversified. In a way the acronym ATBM as such is already indicative of the complexity of the subject. The label most frequently used is "anti-tactical ballistic missile," but it is argued that this is

---

2 Paul B. Stares, *Space and National Security* (Washington, DC: Brookings Institution, 1987); Kenneth N. Luongo and W. Thomas Wander, The Search for Security in Space (Ithaca: Cornell University Press, 1989).

misleading because the *T* in ATBM could stand for theater as well as tactical. The latter interpretation covers both missiles that might be used on the immediate battlefield (battlefield missiles) as well as those that might be launched from areas outside the battlefield against targets within the battlefield (theater missiles). The ballistic variety of battlefield missiles belong to the category of tactical ballistic missiles that have relatively short ranges (below 1,000 km). Theater ballistic missiles have ranges greater than 1,000 km, but shorter than those regarded as strategic (i.e., ranges less than 5,500 km). This may lead to the conclusion that the term "nonstrategic" would refer to the ballistic threat primarily to be countered by ATBMs and it would provide an adequate basis for defining the issue. A distinction between strategic and nonstrategic missiles was indeed already defined in the INF Treaty and SALT Accords (and at least practiced in the ABM Treaty). Unfortunately this definition is of legal character having little to do with the technical and military realities of the potentials of weapons involved. According to the provisions already established, ballistic missiles are nonstrategic if they are land based[3] and have a range of less than 5,500 km, or submarine-based and have a range of less than 1,400 km.

Obviously, a definition of ATBMs using the term "nonstrategic" cannot avoid the disadvantage of drawing artificial boundaries that do not correspond with all the weapon potentials contributing to the regional missile challenge. It would not consider, for instance, long-range strategic ballistic missiles adaptable to shorter distance target coverage (e.g., variable-range ICBMs). The same applies to the broad spectrum of air-breathing missiles, which may include long-range cruise missiles as well as stand-off air-to-surface missiles and other air-launched weapons delivered by high-performance aircraft. The particular defense requirements posed by the weapons have prompted some experts to strike the *B* form ATBM and to place it in brackets or to use two acronyms ATM/ATBM. It was felt that this would better correspond to the realities of air defense and missile defense technology. As an example, upgraded ground-based air defense missile systems capable of intercepting sophisticated air targets should be mentioned, which always include a residual ATBM capability. Development of space weapons may emerge from the potential residual capabilities of ATBM and air defense systems.

Neglecting these complexities, the original approach to an ATBM defense for Western Europe emphasized ballistic missile defense. The interest centered on the nuclear threat posed by ballistic missiles, especially by those of the longer range category, which should be encountered in the long run by a comprehensive theater defense system. In practice, it was only a possibility to start with defenses against theater ballistic missiles of the shorter range category. Improvements of these capabilities were thought to eventually result in a defense against the longer range categories (e.g., by new generation weapon for terminal defense).

Western observers also saw evidence of a similar program in the Soviet Union, which was viewed to justify a particular response of NATO. Reports attributed to the Soviet SA-X-

---

3 Land-based ballistic missiles of the above range category are also referred to as Intermediate-Range-Nuclear-Forces in the INF Treaty. At the same time the distinction between longer range INF systems (range 1,000-1,500 km) and shorter range INF systems (range 500 - 1,000 km) was made.

12 surface-to-air missile having some anti-"tactical" ballistic missile capabilities and, in addition to it, a marginal capability to intercept longer range theater ballistic missiles and even strategic ballistic missiles.

Prior to the 1987 INF Treaty, Western defense planners felt particularly challenged by the triple warhead SS-20 theater missiles (range: 5,000 km) and the single-warhead SS-21, SS-23, and SS-22 theater/tactical missiles (ranges: 120 km, 500 km, and 900 km). Various types of sea-based ballistic missiles added to this threat potential. The Soviet Union had to consider primarily the US/NATO Pershing II (range: 1,800 km) and Pershing IA (range: 720 km), the submarine-based British Polaris A-3 (range: 4,600 km), the French missile forces not assigned to NATO (SSBS M-20 missiles, range: 3,000 km; SSBS-S3 missiles, range: 3,500 km; sea-based MSBS M-4 missiles, range: 4,000 km and, of course, the Chinese missile potential.

All the types of ballistic missiles listed above had been designed to carry warheads of the nuclear variety. Ballistic missiles of the lower range were dual capable (nuclear and conventional) or even triple capable (nuclear, conventional, and chemical). Due to the destructive effects of the particular weapon systems, nuclear ballistic missiles -- especially in their longer range varieties -- clearly posed the most demanding requirements, whereas defense against the non-nuclear missiles categories was comparatively less challenging. For this reason, any attempt to design a comprehensive defense covering the whole variety of ballistic threats would also have resulted in a requirement to implement the component of a strategic BMD system. This could have been accomplished only by a concept that also incorporated space-based weapon systems of a layered ballistic missile defense as envisaged by the U.S. SDI concept. In an effort not to duplicate SDI, planners in Western Europe eventually found convincing political, military, budgetary, and not at least technical reasons against a regional off-shoot of a global strategic defense system.[4]

Taking into account such factors as countermeasures made it highly questionable whether striving for a reliable defense against the nuclear variety of the ballistic threat would ultimately pay off. Moreover, another challenge surfaced that was felt to acquire an immediate response and that seemed to be better suited for being countered by available technologies. Fears centered around a perceived Soviet capability to trigger the collapse of NATO's defense in Europe without crossing the "nuclear threshold" by simply resorting to non-nuclear ballistic missile arsenals, which apparently had been modernized at the time. Many saw evidence of a Soviet incentive to also equip SS-21s, SS-22s, and SS-23s with precision-guided conventional warheads, smart submunitions, fuel-air explosives, or modern chemical warheads. Another modernization program pertaining to cruise and other air-breathing missiles was felt to serve the same purpose. Conversely, the Soviet Union for its part might have created an incentive to react on Western ballistic and cruise missile programs and on the Follow-On Forces Attack (FOFA) concept of NATO designed primarily for

---

[4] Manfred Wörner, "A Missile Defense for NATO Europe," *Strategic Review* (Winter 1986), pp. 13 - 20; Uwe Nerlich, "Missile Defenses: Strategic and Tactical," *Survival* (May/June 1985), pp. 119 - 127.

conventional warhead delivery. The Soviet Union in fact introduced upgraded air defense systems, but not much is known about whether NATO programs also had been relevant for a Soviet decision to improve ATBM defense capabilities.

At least on NATO's side the issue of a comprehensive regional ballistic missile defense was brought down to the concept of "extended air defense,"[5] which should establish a supplementary defense against lower range ballistic missiles by exploiting the capabilities of upgraded ground-based air defense missile systems. Priority had been given to a defense against non-nuclear missiles, whether ballistic, semi-ballistic, or cruise, whether ground-launched, air-launched, or sea-launched, including all the types of manned and unmanned aircraft. In its ATBM part this concept was rather limited, since defense should focus on the Soviet SS-21s, SS-22s, and SS-23s, which had been identified as rather capable carriers of non-nuclear warheads.

When the 1987 INF Treaty was signed, this, of course, resulted in a major modernization of perceptions. At the time, it became clear that a considerable portion of the INF forces, namely, ballistic missiles with ranges between 500 and 5,500 km would be dismantled. In the meantime fundamental changes in East-West relations took place, which would result in a further reduction of weapons. In this context, also the prospects for reaching a supplementary agreement concerning short-range ballistic missiles with ranges below 500 km have in fact substantially improved. In the near future, additional limitations may be obtained for air forces especially for ground-attack versions of aircraft.

Unless such treaties could be achieved, considerable arsenals of nuclear and non-nuclear weapon systems will remain. Presently the arsenals on both sides contain comparatively large numbers of short-range missiles, such as the Soviet Forg-7/SS-21 (range of SS-21: 120 km) and SS-1C/Scud B (range of Scud B: 300 km) as well as somewhat smaller numbers of US/NATO Lance (range: 130 km) and French Pluton and Hades missiles (ranges: 120 km and 480 km).

Future agreements on reductions pertaining to these missile arsenals may further weaken the arguments in favor of an ATBM defense in Europe. But this should not lead to the conclusion that, in due course with the removal of these missiles, the category of regional missile defenses would disappear as a principal option. Moreover, such reductions do not include strategic missiles and the large number of air-delivered systems. Technology upgrades of air defense systems that have a residual ATBM capability may thus persist all the more because these are not subject to immediate arms reductions. It may add to this incentives to introduce dedicated ATBM systems, either derived from BDM or air defense. This arises outside the treaty area where a particular missile threat would materialize. To say the least, the requirements of arms control will keep alive a principal interest in ballistic missile defenses including those having a capability to intercept shorter range missiles. For

[5] Hubertus Feigl, "Tactical Missile Defence," in Walther Stützle, Bhupendra Jasani, and Regina Cowen, *To Defend or Not Defend?* (Stockholm: SIPRI, 1987), pp. 154 - 73.

the foreseeable future, this would also keep open the question of residual space applications of such weapon systems.

III - Operational Requirements of ATBMs

Whether an ATBM/air defense system can be used as a space weapon depends on the operational features of the targets to be intercepted and the characteristics of the particular kind of weapons used for this purpose. The requirements of intercepting ATBM targets (air targets) for an ATBM would thus greatly determine which kinds of ATBM systems (air defense systems) have to be considered as a potential space weapon and the extent to which target engagement in outer space can be achieved by these weapons.

The identification of space weapon applications is comparatively easy if the use of space-based weapons is involved and/or ballistic targets do exist that, while traversing outer space, can be intercepted. Interactions for that kind may significantly differ from engagement scenarios confined to ATBM intercepts within air space. It should be considered in this context that the potential space weapon options of a particular ATBM weapon are ultimately much more determined by the less demanding ASAT requirements that may even be met by weapons having limited capabilities against ballistic targets.

The operational requirements of an effective ballistic missile defense in Western Europe may help to understand the problems of identifying potential space weapon uses of ATBMs. Regional missile defense systems have to counter large numbers of ballistic missiles belonging to different range categories. This causes a variety of ballistic trajectories with flight parameters ranging from the depressed to the lofted variety and reentry vehicles characterized by a wide range of reentry speeds and angles of reentry. Since the speed of a missiles varies with its range, the operational requirements to be met by a defense may differ considerably. The comparatively low reentry speeds of tactical missiles would indeed pose less stringent defense requirements than the much higher reentry speeds of strategic missiles. In contrast, the reduced overall flight times of a missile at lower ranges may considerably reduce the time intervals available for intercepts. Flight times as short as 3 - 4 minutes for short-range missiles (e.g., for SS-21s) may in fact overload the reactive capacity of a missile defense. For a terminal defense the resulting time squeeze may outweigh the advantage originating from the lower approach speeds of these ballistic targets.

Important characteristics of ballistic missiles and the reentry vehicles, both strategic and nonstrategic, are compiled in Table 7.1.

Table 7.1 Missile and Rentry Vehicle Characteristics[a]

Range in km	Reentry Speed	Reentry Angle degs	Apogee in km	Missile
10,000	7.2	22.6	1,325	Minuteman II
5,000	5.9	33.8	988	SS-20
4,000	5.4	36.0	813	Poseidon
2,500	4.5	39.4	560	SS-N-6
1,800	3.9	40.9	409	Pershing II
900	2.9	43.0	222	SS-12
500	2.9	43.9	120	SS-23
120	1.1	44.7	30	SS-21/Lance[b]

a = These figures assume minimum energy trajectories in a vacuum. Atmospheric drag encountered on reentry will reduce a missile reentry vehicle's speed to a greater or lesser degree, depending on its aerodynamic characteristics.

b = The missile's trajectory is so low that it does not exit the atmosphere.

Depending on the particular range category of ballistic missiles the apogees of ballistic trajectories may vary significantly. If referred to an air-space boundary (e.g. 100 km) the data listed in the table would also indicate the extend to which these trajectories run through outer space. Quite in contrast to the trajectories of long-range ICBMs (e.g. Minuteman II), SLBMs (e.g. Poseidon), and intermediate range missiles (e.g. SS-20) which are widely known as space trajectories those of lower range missiles may lie largely (e.g. SS-12s, apogee 220 km, and SS-23, apogee 120 km) or even entirely (e.g. SS-21 or Lance, apogee 30 km) within the atmosphere. Also depending on the particular flight parameters of these missiles, the time intervals exploitable for intercepts in outer space would vary considerably. Whereas ICBMs may travel for more than 20 minutes outside the atmosphere, the respective time interval for theater missiles can be much shorter.

The respective flight parameters of a missile have an overriding importance for selecting a particular defense concept and for optimizing technical solutions according to the requirements (e.g., the relation between intercept ranges and weapon basing). The requirements of a comprehensive approach could obviously be met by a layered ballistic missile defense, as envisaged in the US SDI program.[6] This concept provides an opportunity to extend intercept missions to all flight phases of a ballistic target and to increase the efficiency of defense by making the target subject to repeated intercepts. Operational parameters of the targets to be intercepted, which may be examplified by low apogee trajectories and reduced flight times of shorter range missiles, may restrict or preclude the introduction of layered missile defense, which would have to resort to space-based weapons for boost-phase and midcourse intercepts. Reduced flight times of ballistic systems may also cause difficulties for space-based sensors needed, for instance, for target acquisition and

[6] Herbert Lin, *New Weapons Technologies and the ABM Treaty* (Washington, DC: Pergamon-Brassey's, International Defense Publishers, Inc., 1988), p. 13.

tracking. Only the missiles of the regional category having longer trajectories (e.g., the SS-20s, apogee about 1,000 km) would allow enough time for space-based weapons, which, in conjunction with space-based sensors, could accomplish exoatmospheric intercepts. Boost-phase options would gain from exotic weapon developments, which eventually may allow for interception of all kinds of ballistic missiles within the atmosphere. But conceivably any near- term solution for regional missile defenses would have to rely on ATBM technologies only capable of terminal and late midcourse intercepts.

IV - Ground-Based ATBM/Air Defense Missile Systems

Placing emphasis on terminal and late midcourse defense would certainly have its operational drawbacks. This is mainly true because it implies that, in the absence of 'exotic' weapon technologies, interception of targets would be carried out from those areas that are to be protected. The requirements of this approach would have to be met by ground-based missile technologies and non-nuclear warhead technologies, etc., which may render only limited capabilities.

Near-term solutions for ATBM systems could only be derived from existing ground-based air defense systems, which allow component upgrades. Systems for intercepting air targets are characterized by ground-based radar components and high-acceleration ground-to-air missiles. Examples of this are provided by modern air defense systems, such as the U.S.-NATO Patriot system and the Soviet SA-10 and SA-12 systems. Radar and missiles components of modern ground-based air defense systems allow intercepts of air targets at all operational altitudes. Systems of that kind may be characterized by a maximum intercept altitude of about 40 km, and maximum ranges (slant ranges), which vary between 70 km (e.g., Patriot) and 300 km (e.g., the Soviet SA-5). Defense technologies could allow coverage of a wide range of high-speed air targets, including those travelling at supersonic speed (maximum speed about Mach 3 or 1,0 km/sec) and altitudes exceeding 30 km.

Ground-based missile systems employed in an air defense role may derive profit from superior speed parameters (e.g., 1.6 - 2.0 km/sec for Patriot).

If the same system were used for ballistic missiles defense, the relation between the speed of an approaching object and the speed of the defense interceptor may be even more important because the speeds of ballistic reentry vehicles may greatly exceed those of air targets. As listed in Table 7.1, the 120 km Lance and SS-21 missiles have a reentry speed of about 1.1 km/sec. The 500 km SS-23 reenters at about 2.2 km/sec and the 900 km SS-12 at about 2.9 km/sec. The speeds for the 1,800 km range Pershing II is about 4 km/sec and that for the 10,000 km range ICBMs about 7 km/sec. The maximum speed of the Patriot interceptor missile lies between 1.6 -2.0 km/sec.

The impact of speed relations on intercept points and other operational requirements is difficult to assess, especially if target engagement occurs within the atmosphere. So an increased acceleration of the interceptor may help to compensate a speed deficit, but determinants of that kind would not change the situation basically as long as only a "traditional" technology is used. Under optimum conditions, an ATBM defense against missiles of the lower range categories would only allow for a maximum intercept range of about 20 km, which no doubt lies well within the atmosphere. Extending target engagement to longer range missile categories having reentry vehicles with much higher approach speeds would result in a further reduction of the maximum intercept ranges.

As warheads need to be intercepted as early as possible, it is of crucial importance to make optimum use of the time available for target engagement. For this purpose, defense radars must accomplish target detection, discrimination, and tracking from as far away as possible to allow time for interceptors to accelerate and to fly to intercept points.

The intercept range achievable against a particular missile type determines the size of the area that can be defended by an ATBM battery. This area decrease considerably when the approach speed of reentry vehicles increases. It is important to realize that even under most favorable conditions, the size of the zone would be rather limited.

There is no doubt that first generation ATBM systems derived from air defense systems could hardly accomplish intercepts of ballistic reentry vehicles outside the airspace. But this observation does not automatically imply that missiles developed for ATBM or air defense purposes would not have a range necessary for leaving the atmosphere. Nevertheless, it might be questionable whether slant ranges of several hundred kilometers attributed to some types of air defense missiles would really result in a meaningful ASAT capability. But it should be considered in this context that intercepting in-orbit satellites would pose less stringent requirements, which could ease the interception task. Preplanning of important target engagement parameters may, for instance, help to make better use of the entire range capacity of a missile.

Technical parameters of missile systems specifically designed to intercept in-orbit satellites at great distance may differ significantly from those characterizing ATBM missile systems. In order to achieve a major ASAT capability based on long-range interceptors, a particular design, not necessarily leading to optimized ATBM versions, would have to be implemented. In this context is should be considered that long-range exoatmospheric intercepts require a missile with moderate acceleration and high burn-out speed, whereas in the more stressful environment below 50 km, intercepts must be performed by high acceleration missiles with lower burn-out speed. This example also shows that a missile achieving more than an almost negligible ASAT capability can be identified by means of the specific design characteristics. Similar examples may be provided by the radar components that would have to be adapted to space mission control.

Ground-based missile interceptors for exoatmospheric ballistic missile defense may nevertheless remain a significant issue. This is mainly due to the fact that terminal defenses can only be effective in a comparatively late phase of a missile attack. This deficiency could be corrected to some extent by a shift of intercept points to the late midcourse phase of incoming ballistic warheads. Thus even ground-based missiles may employ shoot-evaluate-shoot techniques, which may result in an increase of efficiency as envisaged in SDI-like layered defenses. For this purpose missiles with specific design characteristics, which differ from those of air defense upgrades, would have to be developed. It is the category of dedicated ATBM systems expressly designed for engaging high-velocity reentry vehicles at greater intercept ranges. This kind of sophisticated ATBMs, which would constitute an exoatmospheric defense layer, could be a derivative of late midcourse intercepts techniques being developed for strategic BMD purposes. The U.S. Exoatmospheric Reentry Vehicle Interceptor Subsystem (ERIS), also capable of midcourse ATBM intercepts, comes to mind, which may be employed in conjunction with an air-based optical sensor system. The infrared-guided interceptor destroys ballistic targets by collision (maximum altitude of intercept more than 120 km, maximum range more than 300 km). ERIS could be supplemented by the High Endoatmospheric Defense Interceptor system (HEDI) with a maximum altitude of intercepts of about 50 km and a maximum range of more than 60 km. The ERIS missile would clearly be capable of ATBM midcourse intercepts. It would also have a limited shoot-evaluate capability against ballistic targets belonging to the longer range missiles categories. It may be important in this context that target kill by collision above the atmosphere is easier to achieve than in the atmosphere. The technology of ERIS is traceable to the Homing Overlay Experiment (HOE), which proved that an ICBM carrying a hit-to-kill warhead can hit an ICBM warhead in near-earth space. HOE also demonstrated that even the existing long-range ICBMs have an inherent BMD capability.

ATBM programs following the technology baseline toward exoatmospheric defense could no doubt render a residual non-nuclear ASAT capability against satellites in the low earth orbit. A more effective operational capability allowing a comprehensive coverage of low-altitude satellites and an extension of intercept missions to satellites at higher altitudes would, of course, require more capable ASAT weapons.

For all ASAT operations involving ATBMs, a space surveillance system would be needed. For this purpose military as well as civil surveillance installations (e.g., radar systems and passive optical sensors) must be linked together with the ATBM sites. In addition, satellite telescopes and sophisticated camera systems could be employed for detection, identification, and tracking of space objects. An operational linkage between space surveillance installations and system components belonging to the command and control structure of ATBM/air defense installations would be highly indicative for the de factor existence of an ASAT capability.

V - Nuclear ASAT Options of ATBMs

Nuclear ASAT options would have several disadvantages, but due to the existence and proven capability of nuclear weapons, such modes of employment cannot be discarded completely in the ATBM context. It is widely known that nuclear warheads, when detonated in space, radiate energy and disperse debris more or less uniformly in all directions. They may upset or damage unhardened satellites at ranges that depend on weapon yield, but that can be very large. Multimegaton nuclear warheads could harm normal satellites hundreds of kilometers away. Omnidirectional destructive power and great lethal ranges would contribute to cover extended space segments, but at the expense of indiscriminate interference with all hostile, friendly, and neutral space objects moving in those segments.

Nuclear explosives can be delivered by a variety of launch vehicles, such an ICBMs, and SLBMs, which, for ASAT missions would need only a moderate guidance accuracy. Above all, the warheads of ICBMs reaching apogees of up to 1,400 km, can be used to intercept a variety of low-orbit satellites.

With respect to the question under consideration, it may be important to know whether nuclear anti-ballistic missiles of the ATBM variety could reach several hundred kilometers altitude. Such missiles could, at least in principle, be used to intercept low-orbit satellites. The use of such a rather powerful and indiscriminate weapon may be highly questionable at the outset. In contrast to ICBMs, SLBMs, and "traditional" ICBMs, the available ATBM interceptor missiles would not lack the kind of precision guidance necessary, for instance, to selectively engage a satellite by a conventional explosion or even by colliding with it. The opportunities provided by the longer range destructiveness of the nuclear-armed ATBM should, therefore, carefully be considered against the considerable risks involved. Nuclear explosions above the atmosphere may cause a number of disruptive effects (nuclear electromagnetic pulse, radar blackout, etc.), which could also harm objects on friendly territory below. Employing weapons with lower yields and shifting the point of explosion to a safe distance would help to overcome some of these difficulties. But for such modifications, long-range missiles, such as ICBMs, SLBMs, and ABMs having the necessary high apogee trajectories, would be much more appropriate. Opportunities of that kind would hardly exist with ground-based or air-borne ATBM missiles restricted by employment parameters, which come close to those of nuclear short-range ABMs. As a result, massive collateral damage in the areas to be protected could hardly be avoided. Corrective measures such as lowering the yield of warheads would only help if exoatmospheric points of explosion several hundred kilometers away could be achieved. Nuclear ATBM options for ballistic target engagement involving significant restrictions of intercept ranges would, therefore, not attain a significant importance. Nuclear ATBM employment against satellites may derive profit from exploiting the full range of interceptor missiles. But whether this would significantly increase the attractiveness of nuclear ASAT option is doubtful especially in view of the omnidirectional weapon effects endangering also friendly satellites.

This assessment has to be qualified possibly to some extent, for nuclear directed-energy weapons such as X-ray lasers becoming available. Such weapons could have a far greater lethal range than the nuclear explosion devices that power them.

Due to the propagation characteristics of X-rays, such weapons could only be employed outside the atmosphere against targets in outer space. X-ray lasers belong to the category of directed-energy weapons that concentrate the radiation effects on limited space segments and minimize unintended damage elsewhere in space or on the ground. This kind of weapons could be delivered by a great variety of launch vehicles and space platforms. The theoretical potential of such devices is not well known today but may make them superior to normal nuclear weapons in the more demanding BMD applications. Thus they may be effective against much simpler ASAT engagements. This assessment would, of course, also pertain to ATBMs, which by their range could attack targets in outer space. It should be considered that the existence of non-nuclear intercept techniques possibly more attractive for countering ballistic warheads may raise considerable doubts concerning the desirability of nuclear-directed energy weapons. Under these circumstances, it can scarcely be expected that the formulation of a nuclear ATBM requirement would work as a development driver for X-ray lasers. In any case, the build-up of a respective capability could never be concealed since extensive testing is indispensable. If developments incorporating ATBM/air defense missile technology would lead to X-ray ASATs, these weapons may not only endanger low-orbit satellites. It should be considered in this context that depending on the particular mode of employment, X-ray weapons could considerably reduce the risk of collateral damage.

VI - Non-Nuclear Directed-Energy Weapons

Directed-energy weapons or beam weapons of the non-nuclear variety have increasingly become the subject of intense research and development.[7] It should be observed in this context that in fact a broad spectrum of potential uses has been covered by these programs. The rising interest in strategic beam weapons for applications such as missile defense and anti-satellite warfare detracted the attention from emerging tactical applications of these technologies. At the present stage it is hardly possible to evaluate the substance and the possible outcome of these developments. This is especially true for the characteristics of new "exotic" ATBM and air defense systems and the potential space weapon uses of these developments.

There are several types of directed-energy weapons that lend themselves obviously to space weapon application. Satellite vulnerability that is higher than that of warheads may again serve as a indicator for this kind of employment option. New ATBM and air defense systems could mainly be built around ground- and air-based lasers and, to a lesser degree, also around space-based lasers. This corresponds to the operational priorities pertinent to ATBM and air defense, which so far, for instance, discarded the space-based alternative. Lasers are ranking high in the priority list of new technologies for ATBM employment. This does not extend to other types of directed-energy weapons. Unlike laser beams, charged particle beams fall apart if they go from the air into space. Neutral particle beams might propagate in space, but the accelerator technology needed to generate them is much further off than anti-satellite lasers. The prospects for microwaves are not yet clear. Their usefulness in electronic warfare to interfere with attacking reentry vehicles is also not clear. Microwave generators may, however, attain importance for air defense at short ranges.

There are several concepts for ground-based laser weapons that could either meet the requirements of an advance air defense system or those of a layered BDM system. Of course, both approaches would imply a capability to interfere with space objects such as satellites. This capability may be residual in case of the battlefield or ship defense lasers, due to the much lower energy levels of mobile or transportable laser devices. Most of these programs are intended to provide for a close-in defense capability against aircraft, helicopters, cruise missiles, and other short-range tactical missiles. But it cannot be excluded that this kind of development would result in a technology base, which eventually would allow for a more effective interference with space objects.

---

7 Jeff Hecht, *Beam Weapons. The Next Arms Race* (New York: Plenum Press, 1984).

Most of the effort goes into the BMD area under the conceptual framework of a layered defense concept. Ground-based laser concepts for direct ground-to-space target engagement would only allow for improved terminal and late midcourse defenses. More refined concepts would render an interception capability that also pertains to the boost-phase segment. The latter variation would comprise a small number of ground sites equipped with laser beam generators, components that provided target acquisition, tracking and pointing capability, and advanced beam control subsystems. In a particular system architecture that may render midcourse as well as boost-phase intercepts, these stations would generate a short-wavelength beam, condition the beam to compensate for atmospheric distortion, and project the beam on to space relay mirrors that redirect the beam to a target. Both concepts would no doubt lead to systems having the attributes of an effective space weapon. Laser weapons in their air-based variety may attain a major importance because they could provide for an effective self-defense for aircraft. But other options more related to space weapon uses are possible in the near future. The use of aircraft as a platform for high-energy lasers would overcome the problem inherent in ground-based lasers. High altitude flights above most of the atmosphere and all of the cloud cover would help to avoid a good deal of the atmospheric beam propagation problems. In addition to this, air-mobile lasers offer more flexible employment options. This extends in particular to interception of satellites in well-known orbits, because an airplane can fly to the right place of attack before a satellite comes into range. But the operation of anti-satellite weapons on aircraft also creates some problems since tracking and positioning systems must compensate precisely for motions of the aircraft, drifts, and vibrations. These difficulties would be aggravated if air-based lasers would be employed against ballistic targets, though there is a principle possibility to engage such targets.

Space-based laser concepts would consist of a constellation of orbiting laser battle stations. This defense architecture may provide in principle a capability of intercepting every satellite in orbit and of all kinds of ballistic missiles launched from anywhere on the earth. The main advantage of these weapons has been seen in their capability to destroy postboost vehicles before all warheads decoys and penetration aids are deployed. Depending on the available technology and cost-effectiveness, these battle stations in space may also contribute to an ATBM defense against ballistic missiles, especially if these would move long enough above the atmosphere. Furthermore, since the beam of some types of space-based lasers would penetrate into the atmosphere down to the cloud top, these weapons might be able to provide some capability against aircraft, cruise missiles, and short-range ballistic missiles, which do not leave the atmosphere. Depending on the atmospheric conditions and the laser wavelength, absorption at altitudes above 10 km may be between 5 percent and 50 percent. At lower altitudes it may vary between 10 percent and 80 percent. Handling the target engagement sequence for endoatmospheric interception from space would, of course, raise difficulties that could hardly be overcome in a cost-efficient way. Though the attractiveness of such modes of employment may be lacking, systems of that kind have to be qualified as space weapons due to the capability to interfere with targets in outer space and in the air space (or on the ground). But requirements of air and missiles defense could well drive a development toward space weapons that have an ASAT capability.

VII - Problems of Distinguishing between ATBM Operations and Space Weapon Operations

When ATBMs may be classified as space weapons is not an easy question to answer. From a technical point of view, ATBMs can be understood as a subcategory of BMD as well as an "offspring" from upgraded air defense.[8] Both varieties of weapons share certain technical characteristics, whereas most of the operational characteristics (e.g., in their ground-based missile versions) remain quite divergent. Functional overlaps thus result through the particular requirements to be met in defending against low flying, highly maneuverable aircraft and air-breathing missiles of the cruise-missile type. These may differ considerably from those for intercepting semi-ballistic air-breathing missiles approaching the target areas from high altitudes and steep angles of attack or those for countering ballistic missiles that attack on fairly rigid trajectories with high reentry speeds and steep angles of reentry. Overlaps do not only exist between air defense and BMD systems but also between BMD and ASAT systems. Certain types of ASAT systems would possess a marginal BMD capability, and many types of BMD systems would have a quite substantial ASAT capability. But also for effectively performing BMD and ASAT systems, the operational requirements may differ greatly.

ASAT is generally a less demanding task than BMD. Orbiting satellites are unarmed and hence pose no direct threat. The incoming ballistic warheads may constitute an immediate threat to satellites because these targets are small in numbers and they follow paths that are know and predictable. There is no time pressure in attacking them, quite in contrast to ballistic missile defense, which must react on the initiative of the offense and handle many targets in a short time. Missiles and warheads move on trajectories that are not known before and at best predictable shortly after launch. Apart from operational characteristics, the strong technical relationship between air defense, BMD, and ASAT prevails in certain modes of employment (ground-, sea-, air- and space-based) and in the respective methods of interference with targets (e.g., nuclear and non-nuclear). The resulting pattern of overlaps may create significant residual capabilities that contribute to the so-called inherent capabilities of a weapon system.

It is this combination of residual capabilities that ultimately complicated the demarcation of particular employment options. This is especially true for ATBM and upgraded air defense systems, which, due to their particular operational capabilities, are characterized in many respects by much tighter limits for space operations. Hence the question concerning the potential role of ATBM systems as space weapons greatly refers to residual capabilities. Thus in formulating a working definition of space weapons, such inherent capabilities of other systems need to be considered.

---

[8] Hubertus M. Feigl, "Tactical Missile Defence," *in* Walther Stützle, Bhupendra Jasani and Regina Cowen (eds), *The ABM Treaty. To Defend or not to Defend.* SIPRI, Strategic Issue Papers (Oxford: Oxford University Press, 1987).

VIII - Summary

Before dealing with the "inherent capabilities" of ATBM systems, one should first have a look at the capabilities of strategic BMD, also known under the acronym ABM. These are dealt with in Chapter 6 in detail. Suffice it to say that, for instance, ground-based BMD systems capable of non-nuclear midcourse intercepts of missiles on high apogee trajectories would render much of the technology applicable to ASAT. More extensive ASAT capabilities may be provided by space-based ABM systems that allow for boost-phase missile intercepts.

Due to the particular requirements of ATBM systems, the respective BMD capabilities of such weapons are reduced in many cases, especially if compared with those of ABM systems. Much in parallel to that, also the inherent ASAT capability of many types of ATBM systems is lower than that of the corresponding types of ABM systems. Ground-based ATBM systems, for instance, which are designed to intercept shorter range ballistic missiles on trajectories of lower apogee, may run below most orbiting and all satellites in the GSO. Space-based ASAT options may hardly emerge from a space deployment of ATBMs without these weapons having a capability to intercept the longer range categories of ATBMs. Satellites for exerting boost-phase and midcourse intercepts against shorter range categories would even face difficulties if "exotic" weapon technologies could be employed.

It may be important for assigning space weapon characteristics to ATBM systems that ASAT requirements would never succeed those of BMD. Maintaining only to some extent, the demanding capability of intercepting ballistic missiles or warheads would in general implicate the existence of an ASAT capability according to the given operational limits. Conversely, if weapon systems are at one's disposal, which are specifically designed to intercept in-orbit targets, this would not automatically imply the availability of a significant BMD potential.

The BMD role of ground-based and space-based ABM systems mentioned above -- the rather remote air-based variety may be neglected -- will clearly correspond to the working definition of space weapons given in Chapter 1, either through the criteria of interfering with the normal functioning of objects and/or the criteria of being stationed in outer space and being capable of hitting space object or objects in the air space or even on the ground. In addition to this criteria, a dedicated space weapon could be applied to all kinds of nuclear and non-nuclear interceptors of these weapon categories (with the exemption of ground-based missile employment for air interception of warheads and in particular of space-to-ground employment options of directed energy weapons). Of course, weapon categories to be classified as dedicated with respect to their BMD role may be labelled as non-dedicated with respect to their ASAT capability.

This set of distinctions also applied to weapons designed for ATBM missions. But due to a significant "inherent capability" of such weapons, some qualifications have to be made. Mission requirements of intercepting shorter range ballistic missiles and certain air targets may indeed show a similarity. In addition, the technical solutions available are characterized by functional overlaps, which may add considerably to a "gray zone" of operational options. It may, in fact, depend on the particular employment parameters whether

ATBM systems must be considered more in the configuration of a dedicated space weapon or that of an air space weapon.

This is especially true for ATBM/air defense systems designed to primarily intercept air targets. Air defense systems that make the optimum use of advanced missile or laser technology, may, irrespectivel of the particular basing mode, attain a significant residual capability of countering ballistic targets. Whereas airborne laser weapons may play a significant role in this context, there is considerable doubt concerning the applicability of future space-based laser weapons for air defense purposes. A questionable capability of engaging air targets -- mainly due to restricted beam propagation -- the high absentee rates of satellites and the existence of effective air defense options (e.g., aircraft with a look-down/shoot-down capability, layered surface-to-air-missile defenses, etc.) may preclude that space-based air defense systems become a major development driver for space weapons. More promising ways of implementing advanced technologies for air defense could, however, result in residual ATBM capabilities that attain the quality of dedicated space weapons.

The BMD variety of ATBM/air defense systems is, of course, primarily designed to intercept ballistic targets that are characterized by the flight parameters of intermediate- and shorter-range ballistic missiles. To the extent that these targets are traversing outer space on ballistic trajectories, they also can be classified as space objects. Consequently, a capability of intercepting these objects while moving outside the airspace would obviously meet the criteria of a space weapon. But under conditions valid for near-term defense solutions, the possibility of assigning this criteria to a particular mode of employment is much more obscured. This is mainly because practical solutions for missile defense, in a certain analogy to air defense, have to capitalize on target intercepts within the atmosphere. Such interceptions may be enforced, as already shown, by ballistic targets not leaving the atmosphere and/or by deficiencies of the particular mode of target engagement. The problems of demarcation involved by this have to be seen quite in contrast to the rather transparent case of intercepting ballistic targets outside the airspace through the employment of in-orbit weapons, such as space-based laser weapons. This particular mode of laser employment may even include a limited capability against ballistic targets (and air targets) while traversing the atmosphere or still dislocated on the ground. It should be doubted, however, whether ATBM requirements would give raise to a defense approach which essentially is more pertinent to strategic BMD requirements. Needless to say that these options, if they would materialize, clearly correspond to the dedicated variety of space weapons.

In the case of ground-based ATBM (and that of the rather remote airborne employment), which is by far the most important kind of intercept in terms of available technology, the interactions to be considered are rather diversified. For ground-based terminal defenses, the possibility of identifying particular weapon uses as those of a space weapon greatly depends on the location of the point of intercept, which, as a function of the particular defense requirements and the available ATBM capability, can be exoatmospheric or endoatmospheric. ATBM systems may thus appear in either the configuration of an airspace weapon or that of a space weapon. ATBMs derived from present and near-term ground-based technology are apparently constrained to the endoatmospheric intercept of warheads, thereby including a very marginal capability against low-orbit satellites. With some qualification the same may apply to the air defense variety, which, from the very outset, is developed for intercepting targets that do not leave the atmosphere. This is apparently different from advanced ballistic missile defenses capable of outer space intercept missions against ballistic targets, and hence all the more orbiting targets. In contrast to these deployment options, the mission characteristics pertinent to endoatmospheric system would, of course, not match with those of the definition of a space weapon. This is only true in respect to the capability of intercepting warheads of ballistic missiles. Ground-based ATBM missiles, as other weapon systems, may have a residual ASAT capability, which obviously would correspond to the "inherent capability" of non-dedicated space weapons.

CHAPTER 8: THE PROBLEM OF NON-DEDICATED SPACE WEAPON SYSTEMS

Paul B. Stares

This Chapter addresses the issue of non-dedicated space weapons, namely, those systems not designed for attacking objects in or from outer space but due, to their inherent physical characteristics, nevertheless posess such a capability. In order to determine how much their presence complicates or undermines the prospects for constraining dedicated space weapons, current and potential types of non-dedicated systems are described and assessed for their military significance. The extent to which existing and potential arms control limitations can address this problem is discussed.

I - Introduction

"Space weapons" appear at first glance to represent a very discrete class of armaments with easily definable characteristics. A closer examination of what could conceivably be classified as a space weapon, however, reveals a less obvious and more inclusive set of systems. In addition to those weapons designed for the express purpose of attacking objects in or from outer space, there are also other systems -- not necessarily even weapons per se -- that due to their inherent physical properties possess the capability to be employed as such. Moreover, many of these systems, which have been termed "non-dedicated" space weapons, exist today unlike the vast majority of the "dedicated" variety.

Should negotiations commence to prohibit or constrain the development of space weapons, the vexing problem of what to do about non-dedicated systems will at some stage have to be confronted. More specifically, their existence threatens to undermine the integrity of agreements limiting the development of "dedicated" space weapons by representing a military meaningful alternative or a loophole that could be exploited to gain one. Their presence, moreover, could complicate the task of monitoring and verifying agreements of this kind. Indeed the unavoidable possession of non-dedicated space weapon capabilities has already been cited by the United States as one reason to question the whole endeavour of space arms control.[1]

The purpose of this Chapter is to determine how much of a problem the issue of non-dedicated space weapons really poses to arms control and explore possible approaches to dealing with it. At the outset, however, it is appropriate to lay out in as systematic way as possible the types of non-dedicated space weapons that are of most interest here.

---

[1] See President Reagan's Report to the Congress of U.S. Policy on ASAT Arms Control (March 31, 1984), p. 5.

II - Non-Dedicated Systems

As noted in the working definition of a space weapon in Chapter 1, design motive is not an adequate criterion for distinguishing this class of weapon systems from others. Thus *any device* with the "inherent capability" to "destroy, damage, or otherwise interfere with the normal functioning of an object or being in outer space" or in the earth environment from outer space, can be considered a space weapon. Non-dedicated space weapon systems, like the dedicated variety, can be classified on the basis of their potential operational mission (e.g., anti-satellite attack, ballistic missile defense, earth-directed strikes); their "kill mechanism" or method of interference (nuclear or non-nuclear); and their mode of deployment (ground, sea, air, space). As becomes evident, however, the most feasible space weapons application for non-dedicated systems appears to be for anti-satellite (ASAT) purposes. It is important to note here that an anti-satellite act need not be confined to attacks against spacecraft that revolve around the earth. Because satellites rely to varying degrees on the support of ground-tracking and control facilities for station keeping and mission-related needs, any means capable of destroying, disabling, or otherwise interfering with these supporting ground stations and their communications links is technically an ASAT weapon.[2] However, in the interests of narrowing the area of concern here to practical proportions, the enormous range of possible ways of attacking earth-based components are excluded from the following discussion. It is also important to note that devices with inherent space weapon capabilities need not be weapons in their own right or even military in nature. Some technical conversion, as a result, may be necessary before they can be employed as a space weapon. A useful approach to categorizing non-dedicated space weapons, therefore, is to distinguish between currently available systems that require little or no conversion before use and those that could be employed only after substantial modification and most probably testing as well.

III - Current Systems

Any ballistic missile capable of delivering a nuclear warhead within sufficient range of a space object to inflict damage or other wise interfere with its normal functioning is technically an ASAT weapon. This would include, therefore, currently deployed ICBMs, SLBMs, and any other nuclear-armed missile system with the requisite reach. Depending on how close a satellite was to a nuclear detonation in space, it could be destroyed or permanently damaged in several ways from the weapon fragments, X-rays, neutrons, microwave and longer wave radiation released in the explosion. The electronic circuitry of unshielded spacecraft is particularly vulnerable to nuclear radiation that can travel untenuated in space over thousands of kilometers. The interaction of the nuclear radiation with the upper atmosphere or the metal surface of the satellite can create a large surge of electric current -- sometimes called an Electromagnetic Pulse (EMP) -- that damages its sensitive electronic systems.[3] Even

---

[2] Paul B. Stares, *Space and National Security* (Washington, DC: Brookings Institution, 1987), p. 74.

[3] Theodore B. Taylor, "Third-Generation Nuclear Weapons," *Scientific American*, 256, no. 4 (April 1987), 30 - 39.

satellites not in direct view of an explosion can later suffer damage from the radiation trapped within the earth's magnetic fields.

Manned reusable launch vehicles such as the U.S. Space Shuttle and the Soviet Buran space vehicle also have the inherent capacity to interfere with objects in orbit. Indeed the Shuttle's potential ability to tamper with foreign satellites and even capture them was evidently a sticking point at the bilateral U.S.-USSR anti-satellite negotiations between 1978 - 79. The Soviet Union has since dropped its opposition to the Space Shuttle (except its use for weapons-related purposes), no doubt because it has developed a counterpart of its own. Finally, space objects, particularly their sensor systems, could also be adversely affected by ground-based lasers developed for non-weapons-related missions such as satellite illumination and radar tracking.

Objects in space could conceivably also be damaged or destroyed from the debris caused by the deliberate destruction of a military satellite that had been maneuvered in their proximity. Many if not all of the Soviet military satellites, for example, appear to possess command destruct devices that are used in the event they malfunction so as to prevent their recovery by foreign powers. The propellent that satellites commonly carry for maneuvering in orbit could also be ignited to achieve the same effect.

Using satellites in this way appears no different to the technique employed by the currently operational Soviet co-orbital ASAT system. In this case an interceptor satellite is maneuvered close to its prey, whereupon it is detonated, sending a cloud of shrapnel to disable the target. There are important differences, however. The Soviet ASAT system has either an active or passive sensor system to help guide it to the target and, moreover, a propulsion system for related maneuvering in space. Satellites do not have the former and normally posess only a limited amount of fuel to move about in space. In low earth orbits where objects typically travel at great speeds -- approximately 8 kms per second -- some form of a guidance system and the ability to make quick and substantial maneuvers is practically indispensable for non-nuclear ASAT attacks since even the slightest error in timing can result in major miss distances. It is for this reason that most experts dismiss the ASAT potential of the Progress Spacecraft used to resupply the Soviet space stations. These vehicles employ a cooperative transponder system to help in the docking process with the space station.[4] At higher altitudes such as the geosynchronous orbit where the velocity of objects is much slower, thereby making interception a less demanding task, it is conceivable that an ostensibly peaceful and harmless satellite could be slowly maneuvered in range of another to cause damage when detonated as described above.

The communication links to and from spacecraft are also vulnerable to interference and disruption from the effects of a nuclear explosion in space or in the upper atmosphere. This would temporarily disrupt communications between satellites and their ground stations

[4] Nicholas N. Johnson, *The Soviet Year in Space, 1984* (Colorado Springs: Teledyne Brown Engineering, 1985), pp. 37 - 38.

by the twin phenomena of signal absorption and scintillation.[5] Any radio transmitter with sufficient power, operating at the appropriate frequency and within line of sight of a satellite, could also be used to jam the "uplink" signals to it. Here a powerful competing signal is transmitted to the receiving antenna. The radio transmitters currently used to communicate with military and civil satellites, therefore, could conceivably be employed in this manner. Other radio frequency devices that have been developed specifically for electronic warfare within the earth environment but that nevertheless meet the above specifications, are likewise potential ASAT devices. Satellites at geosynchronous altitudes, which are in view of large sections of the earth's surface, are particularly exposed to this form of attack.

If they are electronically accessible, satellites can also be tampered with by transmitting false commands and information to them. "Spoofing," as this technique is commonly called, might entail, for example, interference with the normal running of a satellite's stabilizing mechanism, thermal controls, or propulsion system.

IV - Potential Systems

Like ballistic missiles, the boosters used regularly to launch satellites and other space systems have the potential to be turned into ASAT devices. Some space launch vehicles, particularly those in operation by the Soviet Union, were originally designed as ballistic missiles, so the addition of a nuclear warhead would not appear to pose a major engineering challenge. Nor would it be very difficult, as is sometimes argued, to jury-rig a conventionally armed device for the same purpose. This could consist of, for example, iron filings or ball bearings appropriately packaged and fused so that when released in the path of a target satellite would cause grievous damage from the resulting high velocity impact. For the reasons described above, however, this is not as easy as it might at first appear. Given the speed at which objects in low earth orbit travel, an extremely accurate and reliable guidance system is essential to bring the warhead within range at precisely the right time or it will miss by a wide margin. The "kill radius" of the warhead could be enlarged to compensate for lower accuracies, but to still get a reasonable probability of intercept the mass of the device would most likely grow to prohibitive levels.[6] The development of such sophisticated homing technologies is not beyond the technological prowess of either superpower as their ASAT and ballistic missile defense program attest. Yet as these programs also illustrate, non-nuclear intercepts in space are not straightforward exercises that can be conducted with high confidence using equipment cobbled together at short notice from other programs. To gain such confidence requires a certain amount of testing that would bear little if any relation to

---

[5] Bruce G. Blair, *Strategic Command and Control: Redefining the Threat* (The Brookings Institution, Washington, DC, 1986), pp. 203 - 4; U.S. Congress Office of Technology Assessment, MX Missile Basing (Washington, DC: USGPO, 1981), pp. 280 - 82.

[6] Donald L. Hafner, "For the Benefit and in the Interests of All: Superpower Space Programmes and the Interests of Third States," in Walter Stutzle, Bhupendra Jasani, and Regina Cowen (eds): *The ABM Treaty, to Defend or Not to Defend?* (Oxford: Oxford University Press, 1987), pp. 197 - 99. I am grateful to both Donald Hafner and Dieter Felske for further information in this regard.

other space activities. This would be extremely difficult if not impossible to hide under the guise of civil-related space research.

The identification of other potential non-dedicated space weapons is more difficult. Conceivably, manned or unmanned space stations like *Salyut/Mir*, could be used for weapons-related activities. The development of laser communication devices might also pose a threat to the satellite sensor systems. Should they be constructed in the future, solar power satellites could possibly be used to interfere with the functioning of other space systems or even to attack targets on the ground.[7]

V - The Significance of Non-Dedicated Space Weapon Systems

The existence of non-dedicated space weapons and related systems poses a problem to arms control to the extent that they offer a meaningful military capability or, alternatively, a covert opportunity to acquire one that cannot be effectively circumscribed or verified by international agreement. The military capabilities of non-dedicated systems can be assessed on the basis of the following criteria:

1. ***Operational readiness***. How soon could the non-dedicated system be readied for use as a space weapon? What does it entail to make it ready? Are trained personnel available to convert it and use it for this purpose? Are the necessary support systems, such as target detection and tracking sensors, also available?
2. ***Target coverage***. What targets does the non-dedicated system realistically threaten? How many such attacks can it carry out?
3. ***Speed of attack***. How quickly can single or multiple attacks using non-dedicated systems be carried out? Is there any warning associated with their use that might allow defensive countermeasures to be implemented?
4. ***Operational confidence***. What is the probability that single and multiple attacks using non-dedicated systems will succeed in their intended mission?
5. ***Operational costs***. What, if any, are the military or political costs associated with the use of non-dedicated systems as space weapons?

With these questions in mind, it is possible to make an initial assessment of the military capabilities of the non-dedicated systems discussed earlier.

Current Systems

Of the non-dedicated systems available for use as space weapons, nuclear-armed ballistic missiles are among the most readily available. Whereas some modification of their guidance and fusing system would be necessary to target objects in space rather than on earth, this

[7] U.S. Congress Office of Technology Assessment, *Solar Power Satellites*, OTA-E-144, (Washington, DC: USGPO, 1981), pp. 172 - 73.

would not be difficult or time-consuming. No additional training to existing personnel would be required to make these changes or to launch the missiles for this purpose. These missiles, furthermore, appear capable of attacking objects in any of the orbits now used by satellites. As the United States and the Soviet Union each possess many thousands of such weapons, multiple targets can be attacked simultaneously or over an extended period. How quickly such attacks can be carried out will depend in part on the altitude of the target. Objects in low earth orbit probably could be attacked in 10 - 20 minutes or less, although some delay may occur before the designated targets pass within range of the launch site. It would take a ballistic missile several hours, however, to reach objects in the geosynchronous orbit. Whereas certain military satellites have been hardened against the long-range effects of nuclear detonations in space, there seems little chance of surviving an explosion that occurs within their immediate vicinity.

Using ballistic missiles in this fashion does entail some very real costs and risks, however. In the first place, there are the opportunity costs of using ICBMs or SLBMs that were formerly assigned to cover other targets. Moreover, unless a nuclear war had already started, the use of nuclear weapons in space obviously runs the risk of precipitating a conflict of this kind. The launch of multiple ballistic missiles against targets in space could also be misconstrued by others as a preemptive intercontinental salvo. The likelihood of this would again depend on prevailing circumstances. Finally, nuclear detonations in space are not discriminating. Friendly satellites either within line of sight of an explosion or that pass through the subsequent radiation may be damaged or destroyed. If this should occur at low enough altitudes, there may also be collateral interference to communications on earth. Their operational utility will depend very much, therefore, on the circumstances of potential use.

As for the U.S. Space Shuttle and the Soviet Buran system, lengthy preparations are required before each flight that obviously limit their operational readiness. Further time is also needed to maneuver the orbiter within range of the target. Satellites in low earth orbit (no more than 600 kms altitude in the case of the Shuttle) are potentially exposed to interference from these two systems, but both the United States and the Soviet Union would need to weigh the risks of such operations against the likely benefits. In particular, they would have to consider the possibility of encountering anti-tampering devices aboard the target space object. Unless the operation can be conducted at some distance to the orbiter, the risk of damage and loss of life may be enough to deter their use in this manner.

Since powerful radio transmitters are regularly used to communicate with satellites, their operational readiness for electronic jamming must be considered to be relatively high. Satellites of all kinds and at all altitudes are potentially vulnerable to interference of this kind although, as noted earlier, they have to be in line of sight of the jammer whether it is on earth or aboard another satellite. One drawback to jamming is that it is difficult to judge in advance how effective such attacks will be. The results are apparently difficult to judge at the time of the attack as well. In comparison to other forms of ASAT warfare, however, electronic interference is probably the least provocative.

Potential Systems

Most of what is said above about ICBMs applies equally to conversion of space launch vehicles to carry nuclear warheads. The one major difference is that space boosters typically require more preparation before launch than ICBMs and other operationally deployed ballistic missiles. More importantly, unless long-range ballistic missiles are banned or drastically reduced in the future, it is unclear what advantage could be gained by converting space launch vehicles for this purpose.

Some of the drawbacks to using nuclear armed ballistic missiles or boosters would be mitigated if they were fitted with conventional warheads. But as argued above, this is not something that can be achieved quickly or without a test program that would run substantial risk of detection. Operational confidence in untested systems has to be considered low when it comes to such a complex activity as satellite interception. The deployment of non-nuclear armed space mines masquerading as satellites in high earth orbits is conceivable but their telltale maneuvering and stationing in close proximity to other space systems would almost certainly raise suspicion as would their deployment in large numbers.

VI - Arms Control and Non-Dedicated Systems

Such as it is, the current space arms control regime, for example the 1963 Limited Test Ban Treaty (LTBT) and the 1967 OST (Articles I and IV respectively), applies equally to non-dedicated space weapons as it does to dedicated systems (for details see Chapters 1 and 2). These agreements prohibit the *use* of such non-dedicated space weapons as radio frequency devices and lasers against certain satellites, but they do not in any way constrain their deployment. In fact none of the systems described earlier are legally prohibited.

Arms control can go some way to restrict the possession and further operational development of non-dedicated space weapons systems, but it cannot preclude it entirely. States in possession of long-range ballistic missiles and sufficiently powerful radio transmitters will always have the inherent capacity to destroy and interfere with the functioning of objects in space. Arms control, however, can attempt to limit the conversion of nuclear-tipped ballistic missiles to conventionally armed and operationally reliable ASAT delivery vehicles. State parties could agree, for instance, to limits on testing ballistic missiles in ways that resemble the operational characteristics of an anti-satellite attack. Commonly agreed parameters of what testing in an "ASAT mode" means, which are not so ambiguous as to create compliance problems at a later date, would need to be defined, therefore, by the negotiators. The same restrictions would also apply to space launch vehicles and the potential conversion of satellites for ASAT use.

Some are sure to argue that it will be possible to test independently and covertly the *components* of an ASAT system under the guise of innocuous space activities. For example, it is often pointed out that the Soviet Union is able to test the launch vehicle of its ASAT system whenever it launches its ocean reconnaissance satellites into orbit even thought it has not evidently conducted an ASAT test since 1982. Gaining operational confidence in the reliability of a particular booster is very different from achieving a trusted ASAT capability, however. Similarly, it is sometimes argued that the maneuvering and homing systems used for civil space rendezvouses and docking provide the basic building blocks for a satellite interception capability. But, as indicated earlier, there are significant and identifiable differences between the maneuvering and docking exhibited in civil space activities and what would be need for ASAT purposes. It is important to point out, moreover, that should covert component testing in a highly disaggregated manner go undetected, there would surely be residual doubts about the effectiveness and reliability of the complete weapon system when combined together and used.[8]

In addition to test constraints, further safeguards could be agreed through such arrangements as minimum separation distances between space objects or "keep out" zones around satellites in high altitude orbits. Given the changing attitudes to on-site inspection, it might be possible and productive to subject space-related facilities to some form of physical inspection to increase confidence in compliance and deter cheating.

VII - Concluding Observations

Certain non-dedicated systems -- notably nuclear-tipped ballistic missiles and electronic jamming devices -- will always pose a threat to the functioning of space objects, although as weapons systems they do have certain operational limitations. The development of potent and reliable non-nuclear ASAT devices under the guise of being innocuous civilian space activities is extremely difficult without a significant test program that risks detection.

In comparison to the types of dedicated space weapons that could be developed and deployed in the absence of formal constraints, the residual threat from non-dedicated systems is minor and need not be a reason to dismiss the practicality of such limits. Dedicated systems pose a much larger threat that would be difficult to defend against unilaterally.

The significance of non-dedicated systems in an arms control regime is related to the nature of the negotiated constraints. Under certain regimes where the testing and development of dedicated types of space weapons is allowed, the presence of residual capabilities may add little if anything to permitted military capabilities. The more comprehensive and stringent the arms control regime, the more significant the existence of non-dedicated systems become. But here again the residual threat from these systems has to be compared against the benefits of constraining more effective types of space weaponry.

The opportunities to convert non-dedicated systems into reliable space weapons can be reduced through testing constraints. Rules of the road for space operations and possibly cooperative procedures to improve the transparency of space activities could further limit the potential for covert development and deployment.

---

[8] Donald L. Hafner, "Negotiating Restraints on Anti-Satellite Weapons: Options and Impact," in Joseph S. Nye, Jr. and James A. Schear, *Seeking Stability in Space: Anti-Satellite Weapons and the Evolving Space Regime* (Lanham, MD: University Press of America, 1987), pp. 101 - 2.

Finally, whereas the opportunities to covertly develop and test component parts of a space weapon system under the guise of permitted space activities can never be reduced entirely, the risk of discovery with all the diplomatic and military consequences that this may bring, is similarly irreducible.

Chapter 9: Definition of Components, Laboratory Testing, and Testing in an ABM and ASAT Mode

Stanislav N. Rodionov

The only arms control treaty that defines components of a weapon system that it controls is the 1972 ABM Treaty. However, this is confined not only to certain types of systems, but it is also a bilateral treaty. The treaty indicates that in defining components, not only is it possible to put some numerical limits to them, but it is necessary to have qualitative limits. In this Chapter, an attempt is made to suggest approaches to the issues of the definition of components and various testing modes to space weapons.

Five components of such weapons have been identified: space platforms, kill mechanism, energy reserve, SATKA (surveillance, acquisition, targeting, and kill assessment), and C^3I. A number of corresponding subcomponents are also defined. In some instances, verification issues are discussed and possible solutions are outlined. Examples of space tests are discussed that might be considered "laboratory testing." With regards to other test issues, components that have substantial military potential and therefore should be banned from testing and deployment are identified.

I - Introduction

The ABM Treaty, the only arms control treaty that defines components of a weapon system that it controls, deals with only the ground-based weapons that are aimed at warheads during the terminal phase of their trajectory. There are other phases of missile trajectories: boosts, post-boost, and midcourse (ballistic). The latter, in principle, does not differ very much from a satellite's trajectory (if, of course, the ballistic phase is outside the atmosphere). The only difference (but a very important one) is that satellite trajectories are predictable with high accuracy, whereas the trajectory of a reentry vehicle could be determined only at rather short notice. One exception is a maneuverable satellite, that is, a satellite that could change its altitude or orbital inclination for a short time compared to its orbital period. The latter could range from 90 to 100 minutes. Another example is the so-called tethered satellite. A tethered satellite is one that orbits the earth linked to a mother spacecraft, and their combined orbit could be changed relatively simply in an unpredictable way.

Thus there are now many more parameters to consider than those mentioned in the ABM Treaty, such as ASAT and antimissile weapons. This is an addition to potential space weapon systems.

The ABM Treaty has indicated that the definition of components not only helps to put some numerical limits on these, but also qualitative limitations are necessary and possible. However, in the case of space weapons, not only the number of possible targets is increased enormously, but the potential of weapons to attack targets has also increased. The definition of space weapons adopted in Chapter 1 includes such categories as air-based systems and even "inherent capability to destroy or otherwise interfere with normal functioning of objects in Outer Space or in the Earth environment."

In this Chapter the issue of the definition of components is considered first. An attempt is made to show what may be considered as laboratory testing. It should be noted that no legal definition of "laboratory testing" is given. The last part of the Chapter deals with various aspects of component testings in space or in near-earth environment.

II - Definition of Components

The ABM Treaty adopted a functional type of definition for components. A similar approach may be relevant in the case of space weapons. Whereas there are many more parameters to be considered now, they may result mainly in the introduction of subcomponents not defined in the ABM Treaty.

In general, one may distinguish between, at least, five components (or elements) of space weapons:

1. Space combat station (which carries weapon system or some of its components).
2. Kill mechanism.
3. Stored energy and system of energy conversion.
4. SATKA system (surveillance, acquisition, targeting, and kill assessment).
5. C^3I system (Command and Control Communication, Intelligence).

In the following sections, these components are discussed in detail as well as possible thresholds for these elements and corresponding verification methods.

Space Combat Stations

These are directly analogous to the launcher mentioned in the ABM Treaty. One may imagine the following subcomponents, which look more or less natural:

- Permanent space platforms (making more than one revolution around the earth).
- Pop-up rockets and/or airplanes.
- Long-lived air-based platforms (for example, aircraft with refuelling or with inherent capability to be refuelled).

Permanent space platforms should have the following features:

- Capability to carry massive and large-size payloads.

- Rapid maneuvering capability, that is, a capability to change to orbital height or inclination in time, comparable to the orbital period of the station round the earth.
- Capability of fast retargeting either by changing orientation of the station itself or by specially designed smaller platforms.

All these features imply that permanent space platforms would have very large amount of stored energy and a great number of engines. These typical signatures may help to identify systems of such a type as a potential subcomponent of a space weapon system.

Depending on the nature of an arms control agreement (bilateral or multilateral), there can be several options for banning deployment in space of this particular subcomponent. For example, if pre-launch inspection would be allowed and an inspecting group would detect in the payload a platform with some suspicious features, inspectors should have the right to inquire about real reasons to launch a system that looks like a combat space station. Moreover, if an explanation of an inspected side were considered unsatisfactory, inspectors should have the right to postpone the launch or even to put a ban on it.

There is a possibility, of course, of constructing a platform not in space by transporting parts that may not seem dangerous to inspectors. In this case the final assembling might be done, for example, by a manned space station crew. However, such an unexpected appearance of weapon's components in space could be detected with a high degree of confidence by the NTM. Such covert activities might have very serious political implications.

It might be possible to put numerical limits on such subcomponents as well as some limitations on deployment zones. This would be analogous to such limits described in the 1972 ABM Treaty. Nobody knows now to what extent politicians can establish "game rules" for limitations in space. From the point of view of full-scale ABM system goals, high orbital inclinations are preferable for combat stations and the GSO for deployment of some SATKA system elements. The altitude for a specific platform depends on the nature of its payload. For example, a platform might be allowed in a near-equatorial orbit. If the total number of such platforms were very low (for example, one or two), then the military potential of such a system would be close to zero for an ABM requirement, whereas an ASAT potential may still remain significant.

The second subcomponent, pop-up systems, seems not to be verifiable before launch. Moreover, its operational time is rather short. The third subcomponent has numerous uses and so there may be only a few features that might indicate its possible use for an ABM or an ASAT. It might be concluded that, from a strategic point of view, space-based platforms are more important than the other two subcomponents (with some probable exceptions for several ASAT cases).

When numerical limits are considered, only those components or subcomponents are relevant that are really identifiable and verifiable. Therefore, only space-based platforms might be considered. Generally, only rather large number of such objects can provide a full-scale ABM system (this system may be considered as a limiting case for deployment of any weapons in space). It may be concluded that: in an hypothetical agreement in which space weapons are limited, there may be room for small number of large-scale platforms deployed in specified space zones. These platforms can have some non-military applications as well.

It should be noted, however, that large-scale, space-based platforms are not the only options for permanent deployment of weapons in space. They are also vital subcomponents for "exotic" weapons based on new physical principles. In contrast, space deployment of rather small rocket-interceptors (such as the U.S. "brilliant pebbles") does not need, in principle, large-size platforms. The most important feature of this approach is a size of corresponding satellite constellation which is observable. A single interceptor might be hidden reliably in the payload of practically any satellite. But the strategic value of such a system is doubtful.

Kill Mechanism

Only one type of weapon is mentioned in the ABM Treaty: single-warhead, ground-based interceptors (this does not mean that no other weapons were discussed). The level of technology in the 1970s implied that the ABM weapons consisted nuclear warheads. Now, techniques of weapons, missiles, and satellites seem to have a new qualitative dimension. This is one reason why kill mechanism should be considered as a separate component of space weapons. One can imagine, at least, the following subcomponents of kill mechanism:

- Nuclear warheads.
- Kinetic kill systems (which provide mechanical destruction of target by collision with a rather high relative speed).
- Powerful lasers in the UV, IR, and visible spectral ranges.
- Relay mirrors for retargeting beams from powerful ground-based lasers that have operating wavelengths corresponding to an "atmospheric transparency window."
- Accelerators for charged particles for producing intense beams of neutral particles (which are not influenced by the geomagnetic field).
- Electromagnetic guns.
- Generators of powerful microwave radiation (to simulate the EMP effects similar to that produced from a nuclear explosion).

This list of subcomponents is by no means exhaustive; it is only a first approximation suitable for further discussion.

Two first subcomponents are associated, as a rule, with small rocket interceptors. All others need large-size, space-based platforms.

It is hard to imagine how the idea of numerical quantitative limits could be applied to this class of subcomponents. It seems that practically all systems belonging to the "kill mechanism" component are typical representatives of the so-called dual-purpose devices that may be designed for civilian uses but have the inherent capability to be weapons. So in this case only qualitative limitations and thresholds seem relevant.

For example, one can propose a threshold on laser mirror size. In a similar way, accelerator length (i.e., its energy) may be limited. Thresholds on the size of EM guns correspond to limitations on the final speed of the projectiles and so on. This kind of qualitative limits gives a possibility to put a threshold (at least, partly) on the so-called kill range, which is one of the most important features of any type of weapon.[1]

But some subcomponents, especially those based on new physical principles, are still large enough to be easily detected and identified. One should not forget the dual nature of these systems. They are supposed to be unique for solving many scientific problems and they are useful in a number of civilian projects. This particular aspect needs a very delicate inspection and, at the same time, a very reliable one.

The military potential of many subcomponents could also be determined by the amount of available energy; this important aspect is discussed later.

Regarding verification, it should be noted that conventional space-based interceptors might be a serious problem. These interceptors are designed to hit a target and destroy it mechanically. The most important characteristic of an interceptor is its final speed relative to launching platform -- this parameter determines both a kill range and a kill efficiency. In practice, speeds of several km/s are achieved. If the mass of the non-nuclear warhead is about 10 kg, as it is planned to be, then it is simple to estimate the size of the interceptor before launch. The result is rather pessimistic in the sense that real interceptors would not be large and therefore might be hidden inside payload elements.

It should be remembered that in a full-scale (from a strategic point of view) constellation of a weapon system, interceptors would be numerous and carried mostly by satellite in a orbit with high orbital inclination. Any attempt to begin deployment of such a constellation might be regarded as an indirect indication of interceptor deployment in space even without detailed pre-launch inspection.

Very serious problems might arise if some third country not limited by the ABM Treaty would deploy in space interceptor-like systems and keep their actions secret. The implication of such an adventure on international security and stability may be very negative.

Numerical limits on space-based interceptors are possible if only the verification procedure is provided. It is supposed that the only solution to the verification issue can be very detailed pre-launch inspection. One way might be to inspect all payload elements before integration or monitor them by X-ray techniques. The important issue is that, inspections should not to be intrusive.

---

[1] S. Rodionov, "Pacification of Space Technologies," *International Affairs*, Moscow, No. 12 (1988), pp. 69 - 77; S. Rodionov, "Definition of Permissible and Restricted Activity in Space," in Proceedings of the 38th Pugwash Conference on Science and World Affairs (Dagomys, USSR, 1988), 1989.

Other subcomponents of a large size may not need any numerical limits on deployment if their kill ranges, for example, are be well below a confirmed threshold. In this case the other approach might be more productive, namely, preannouncement about planned mission, experiment schedule, device characteristics, energy, and fuel consumption as well as about recommendations on corresponding verification measures. Of course, such an approach needs corresponding political environment.

Stored Energy

This component is not covered by the ABM Treaty. The reason for singling this out is a straightforward because since the stored energy (or the fuel mass) determines, all being equal, a real kill range of a particular type of weapon. This is true also for interceptors mentioned in the ABM Treaty. But their kill range was predetermined as well as their size.

It is possible to make estimate energy and fuel consumptions for various subcomponents of kill mechanism systems. It has been estimated that every shot of laser weapon with a kill range exceeding 1,000 km would need about 100 MJ of the beam energy.[2] It is well known that laser efficiency in many cases is rather low (of the order of a few percent). So the primary energy expenditure would exceed 1 GJ per each shot.

A projectile of 10 kg mass with a speed of 5 km/s has a kinetic energy about 100 MJ. Such a projectile may be accelerated either by a rocket interceptors or by an EM-gun. In either case the efficiency would be of the order of 10 percent, so that a primary energy reserve should be close to 1 GJ again.

It was mentioned earlier that space-based platforms should be maneuverable. This capability implies a huge energy consumption. One needs about 5 MJ (tonne. km)$^{-1}$ for a change of orbital height. The change in orbital inclination is even more energy-consuming: it would need about 1 GJ per each tonne of platform mass to change the inclination of the orbit by a single degree.

For many weapon systems, it is useful to store energy (electrical energy, in particular). There are a number of methods available, for example, chemical batteries, capacitors, magnetic storage systems, and flywheels. The primary energy source may be solar energy, chemical fuel, and nuclear energy.

The solar energy flux in space is about 1.4 kw.m^{-2}. Photovoltic conversion has an efficiency of 10 - 15 percent for production of electricity. A new technique, the so-called solar dynamic system, is based on a gas-driven engine (closed Brayton cycle) and solar energy concentrator. It may be possible to obtain up to 100 kWe from the solar light in near-earth orbits.

---

[2] E. Velikhov, A. Kokoshin and R. Sagdeev (eds): *Weaponry in Space: the Dilemma of Security* (Moscow: MIR, 1986).

Chemical fuels for rocket engines deliver about 4 MJ per kg. The same energy/mass ratio is typical for high explosives (TNT, for example). The highest energy output can be obtained by burning beryllium in fluorine atmosphere. About 30 MJ can be generated for each kg of reaction mixture. The energy output of a hydrogen fluoride chemical laser would be only about 0.1 MJ per kg of reacting mixture.[3]

Nuclear energy is often considered to be the most useful energy source for space applications. But thermal power of space-based nuclear reactors should be extremely high if these reactors are designed for strategically important weapons. For space-based nuclear reactors on board permanent platform limiting power/mass ratio is estimated to be 3 - 5 kW/kg[4] (this limit is far above the present operational level). An efficiency of thermal to electrical energy conversion in this case is about 5 percent (the most optimistic estimation is 10 percent). In the more distant perspective, the situation may be changed. The American Physical Society study[5] mentioned space-based pulsed nuclear reactors with the power/mass ratio as high as 0.1 - 1 MW/kg.

It should be noted that all systems of energy sources are rather massive. For example, chemical lithium batteries can have energy to mass ratio as high as 10 MJ/kg but they are very time-inertial with mean specific power of only about 1 kW/kg. Capacitors do not have high energy/mass ratio -- only about 10 kJ/kg, and for magnetic storage units this parameter may be of the order of magnitude higher. Theoretical limit for flywheels is estimated to be in the range of MJ's/kg, but efficiency of mechanical to electrical energy conversion is not more than 1/3.

All conversion systems do not have high efficiency so that they would emit extra heat in space. The emitters need to have huge surface areas that may be considered as specific signatures of the energy system (as well as extra temperature of these emitters). Such features can not practically be hidden from observation and identification.

This is an option to deploy in space autonomous energy modules rather then substantial energy and/or fuel reserve on other space-based objects. This approach might be helpful in putting some limits on the number of energy modules. The other possible approach is connected with preannouncement about energy and/or fuel reserve for any particular mission and about the program of its expenditure.

SATKA System

The SATKA system is a vital part of any space weapon. It has some specific elements, namely, different radars, active and passive laser locators and range fingers, high-speed computers -- space- and ground-based with high operational and logical capabilities and huge

---

[3] Ibid.

[4] *Theory and calculations of Spacecraft's Energy Systems* (in Russian) (Moscow: Mashinostroyeniye Publishers, 1984).

[5] *Review of Modern Physics*, vol. 59, no. 3, part 2, 1987.

memory reserve. Verification of SATKA means detection of a great number of mirrors and antennas, which in some cases may be very large.

Functionally, one may define several subcomponents deployed in different space zones:

- Early warning system (passive IR-sensors deployed in high-altitude orbits including the GSO).
- Boost-phase tracking system (SWIR -- short wavelength infra-red sensors deployed in orbits at altitude above 1,000 km).
- Ballistic-phase tracking system (LWIR -- long wavelength infra-red sensors).
- Active and passive discrimination systems (for distinguishing between real targets and decoys).
- Terminal-phase tracking system (it may be air-based LWIR).

- Satellite tracking and control system (both ground-based and space-based).
- Targeting and pointing system (located very close to corresponding kill mechanism subcomponents).

The SATKA may be analogous to ground-based radars mentioned in the ABM Treaty. If one is going to follow this analogy, then some quantitative limits should be worked out with a possible future treaty limiting space weapons.

It should be noted that in some specific cases qualitative limits may also be given. One particular example is connected with SWIRs, which track the initial trajectories of boosters. In order to identify each single booster, a sensor should have linear resolution in its focal plane of the order of few meters (that is, booster plume size). It is known that linear resolution is equal to a distance between the sensor and the target multiplied by an angular resolution.

For any optical system this resolution is the ratio of operating wavelength to mirror diameter. The typical temperature of missile engine plume corresponds to maximum radiation at the wavelength of few microns. Taking these into consideration, the following simple equation is obtained:

$$R = 10^6 D,$$

where R is the operating range for the sensor and D is the size of the mirror of the sensor. An operating range of 1,000 km would correspond to mirror diameter. In practice, this equations may be used to put some limitations on SWIR operational ranges. Similar equations might be used for setting limits on targeting and pointing system.

C^3I System

This system is very important for a wide network of combat platforms and elements of SATKA. But for a single weapon-like object deployed in space, this component makes no real sense.

The most important part of the C^3I is the extremely sophisticated software used for processing of all input information coming from other components in real time as well as for providing a prompt response in the form of operative instructions for all other components. Any activity of the C^3I component may be detected (at least theoretically) by observation of a sharp increase of information flow between all elements of a space weapon system and a small number of command centers. One cannot predict now to what extent communication line (radiochannels, laser channels, optical fibre lines) might be controlled by an adversary.

The above list of definitions of possible elements of space weapons or weapon-like objects is by no means complete. However, it gives some examples of the way in which numerical and qualitative limits could be drawn out to control deployment of space weapons and related systems.

III - Laboratory Testing

It is easier to define "laboratory testing" for ground based systems, for example, by introducing a new term "underroof testing." This test category is considered as absolutely nonrestricted for ground basing. The problem is how to find similar conditions in space. Three possible approaches seem to be relevant to this issue, but no attempt is made to formulate a legal definition of laboratory testing in space.

The first approach is rather straightforward. It is literally similar to the "underroof" concept. Any test inside a closed volume located on a manned or unmanned space platform could be regarded as a laboratory. In a more restrictive proposal, a condition might be complete absence of any radiation and/or particulate inflow and outflow connected with a test. Under these assumptions, a laboratory test seems to be absolutely invisible to the other side.

The second approach concerns tethered satellites. One of them may serve as a weapon platform; the other may represent a target. Under such a geometry it is possible to fulfil the above mentioned condition, that is, "no inflow, no outflow," although a system is not completely closed itself. Testing under such conditions might be quite safe for any other space object. Chances for some king of external control of this test seem to be rather small but not equal to zero. Possible restrictions on this kind of testing may include limits on a tether length as well as limits on stored energy.

The third approach is more permissive in nature but requires some political preconditions. During the discussions on a possible ban on ASAT activities, one proposal was to set a limit on the distance between two objects in orbit (with some exception for space station docking). It was proposed that this distance could be few tens of km.

If this concept of the closest approach distance (CAD) were to be acceptable, an idea of a "security zone" around any space object (e.g., a sphere with radius equal to the CAD) might be considered. Within the "security zone," some testing might be permitted with certain limitations on the flow of radiation flux and particle beams. Such an approach does coincide with a "kill range" approach mentioned earlier. In other words, the CAD value should be well above the kill range of any system being tested inside a "security zone."

These examples show that the laboratory testing concept has some sense for space-based systems, but solution of this issue depends upon political environment.

IV - Testing in Space

The next logical move from the laboratory testing issue is connected with a problem of "field tests" for space-based systems. Recently some aspects of this problem have been investigated.[6] In particular, a proposal to create an Agreed Orbiting Test Range (AOTR) analogous to the ABM Treaty's agreed ground test ranges has been made, and location, scope, and activities of the AOTR are discussed in more detail.

Another aspect of the "field testing" is the nature of the test itself. We identify those testing activities that encourage development of military capabilities of a given device or a system. However, owing to the complexities of the problems, it is not possible to propose solutions to every corresponding issues in every case.

Consider the following typical situations when only a single component is under test.

Space Platforms

If a large platform with a high load capability and maneuverability is launched secretly (without any explanations or details of mission goals and schedules), there would be considerable concern from major space users. Such a platform may have on board important military payloads launched rapidly by modern launch systems having lift capabilities as high as 100 tonnes. Thus military space platforms may be a source of political instability.

An obvious first response would be to ban such launches. However, a more permissive option might be to limit the number of such platforms in those orbits or space zones that do not provide substantial military potential for such systems. The last, more open option would be total program *glasnost*, open to any control and inspection, and wide international cooperation.

---

[6] A. Carter, "Limitations and Allowances for Space-Based Weapons," in *Defending Deterrence*, A. Chayes and P. Doty, (eds) (Washington: Pergamon-Brassey's International Defense Publisher, 1989).

But it is impossible to identify the nature of any test for two other subcomponents (pop-up and air-based systems) without detailed information on their payloads.

It should be noted that SATKA-dedicated satellites do not have signatures specific for large space platforms.

Kill Mechanism

It should be remembered that such systems are generally dual-purpose. Thus a more efficient way of controlling these systems seems to be qualitative and thresholds limits. These limits should guarantee that a given system cannot reach strategically important military potential even with an infinite capacity of energy sources. Examples of such limitations are given earlier (e.g., limits on laser mirror, on EM gun length, etc.). With such limits, these systems (based on new physical principles) may be used in many civilian programs. However, such civilian activities may be regarded as a means for testing some subcomponents as well.

Tests of some other subcomponents need more serious considerations. For example, testing of any nuclear explosion is banned in the air, under water, and in space. Deployment of nuclear warheads in space is also banned.

The ABM Treaty bans testing and deployment of interceptors in space. However, ground-based and air-launched interceptor tests in ASAT mode are considered as permissive. There are some proposals to permit even testing of interceptors in space if some parameters, such as final speed relative to launching satellite, are limited. This kind of limitation is similar, in principle, to the "kill range" approach.

Energy System

It seems reasonable to ban deployment of all autonomous energy sources in space, and activitics connected with an energy transmission between space-based objects. The only exception may be given to widcly open international programs. Deployment of large solar batteries (at present only theoretical) in the GSO is an cxample.

Every space program with extremely high energy and/or fuel reserve not explained by official declaration should be put under international inspection.

SATKA System

Any test of SATKA subcomponents and elements needs a corresponding target in space. These targets may be naturally occurring or they may be satellites already in orbits performing others tasks. The ABM Treaty allows some test launches from agreed test ranges. Thus, experimental launches of new strategical missiles may represent an other source of targets. Even launches of civilian satellites may be used for some SATKA-related testings. Therefore, it is not easy to put limits on tests of SATKA subcomponents.

It may be proposed to have some qualitative thresholds or limitation in order to reduce operational ranges of active and passive sensors. Some examples of this type were given earlier. The size of corresponding mirrors and antennae as well as energy reserve might be objects of such limitations.

C^3I System

Large-scale testings of C^3I systems are impossible per se without some other weapon components. The most appropriate case is a test with the SATKA system deployed almost at full scale. However, as the full-scale deployment of the SATKA system is banned, the question of testing does not arise. In principle, in order to test the C^3I system properly, it needs to simulate attacks and counterattacks by space weapons. But this kind of testing is rather provocative and, therefore, it should be banned.

Now let us consider situations when pairs of components are included in testing. The total number of double combination among five variables is 10. But not all of then have the so-called synergetic effect (i.e., not in every case the presence of a second component would give a profound qualitative effect from the point of view of military importance of particular testing).

Table 9.1 Relationships among Various Components of a Space Weapons

Components	A	B	C	D	E
1. Platforms	-		d		
2. Kill mechanism		-	d	d	
3. Energy			-		
4. SATKA				-	d
5. C^3I					-

We denote by the label *d* in the Table 9.1 those component pairs for which tests are dangerous from the point of view of providing a military potential to a given system. Tests of these pairs should be banned or restricted. For other pairs possible restrictions are connected with limitations on single component testings.

The total number of triple component's combinations is 10. Tests of those combinations that include "dangerous" pairs are considered as dangerous also. Formally only one triple combination, namely, ABE -- platform + kill-mechanism + C^3I, seems to make no sense for testing. All other combinations should be banned for deployment and testing.

All five quadruple combinations of components are militarily important, and their deployment (as well as testing) should be banned.

Of course, the combination of all five components is a real weapon in space, which should absolutely be banned.

VI - Conclusions

In this Chapter, possible approaches are suggested to the delicate issues of the definition of components and various modes of testing of space weapons.

Five components of space weapons are proposed on the basis of functional definition: space platforms, kill mechanism, energy reserve, SATKA and C^3I systems. Many corresponding subcomponents are defined as well. In some cases verification issues are touched upon and possible solutions outlined.

Three examples of space tests are presented that might be considered as "laboratory testings." This is a rather delicate problem that may be resolved only by political agreement.

More general considerations of space tests issues is concentrated on the identification of those component's combinations that have substantial military potential and therefore should be banned both for deployment and testing.

List of Abbreviations

ABM	Anti-ballistic Missile
AD	Air Defence
AFsatcom	Air Force Satellite Communications
AOTR	Agreed Orbiting Test Range
ARABSAT	Arab Satellite
ASAT	Anti-satellite
ATM	Anti-tactical Missile
ATBM	Anti-tactical Ballistic Missile
AUSSAT	Australian Missile Defense
BMD	Ballistic Missile Defense
CAD	Closest Approach Distance
CD	Committee on Disarmament
C^3I	Command, Control, Communication, and Intelligence
COMINT	Communications Intelligence
COPUOS	Committee On the Peaceful Uses of Outer Space
DBS	Direct Broadcasting Satellite
DEW	Directed Energy Weapon
DMSP	Defense Meteorological Support Program Satellite
DSCS	Defense Satellite Communications System
DST	Defense and Space Talks
ELINT	Electronic Intelligence
EM	Electronic-Magnetic
EMP	Electro-Magnetic Pulse
ENMOD	Convention on the prohibition of military or any other hostile use of environmental modification techniques
EORSAT	Electronic Ocean Reconnaissance Satellite
ERIS	Exoatmospheric Reentry vehicle Interceptor Subsystem
EAS	European Space Agency
EUTELSAT	European Telecommunications Satcllite
FEL	Free Electron Laser
Fleetsatcom	Fleet SATellite COMmunications
FOBS	Fractional Orbital Bombardment System
FOFA	Follow-On Forces Attack
GLONASS	GLObal NAvigation Satellite System
GMS	Geostationary Meteorological Satellite
GOES	Geostationary Operational Environmental Satellite
GPS	Global Positioning System
GSO	GeoStartionary Orbit
HEDI	High Endo-atmospheric Defense Interceptor
HEL	High Energy Laser
HOE	Homing Overlay Experiment
IAEA	International Atomic Energy Agency
ICBM	Intercontinental Ballistic Missile

INF	Intermediate Range Nuclear Forces
INMARSAT	Intermediate Maritime Satellite
INSAT	Indian Satellite
INTELSAT	International Telecommunications Satellite
IRS	Indian Remote Sensing Satellite
ISMA	International Satellite Monitoring Agency
ITU	International Telecommunications Union
JERS	Japanese Earth Resources Satellite
KEW	Kinetic Energy Weapon
KH	Key Hole
Laser	Light Amplification by Stimulated Emission of Radiation
LEO	Low Earth Orbit
LTBT	Limited Test Ban Treaty
LWIR	Long Wavelength Infrared
METEOSAT	Meteorological Satellite
MHV	Miniature Homing Vehicle
MOS	Maritime Observation Satellite
NASA	National Aeronautics and Space Administration (U.S.)
NOAA	National Oceanic and Atmospheric Administration
NPB	Neutral Particle Beam
NTM	National Technical Means
OST	Outer Space Treaty
PBW	Particle Beam Weapon
PTBT	Partial Test Ban Treaty
RAE	Royal Aerospace Establishment (UK)
RFW	Radio Frequency Weapon
RORSAT	Radar Ocean Reconnaissance Satellite
RSMA	Regional Satellite Monitoring Agency
RV	Reentry vehicle
SATKA	Surveillance, Acquisition, Targeting, and Kill Assessment
SBI	Space-Based Missile Interceptors
SBL	Space-Based Laser
SDI	Strategic Defense Initiative
SDS	Satellite Data System
SHF	Super High Frequency
SIGINT	Signal Intelligence
SLBM	Submarine-Launched Ballistic Missile
SPOT	Système Probatoire d'Observation de la Terre
SSTS	Space Surveillance and Tracking System
SWIR	Short Wavelength Infrared
TELINT	Telemetry Intelligence
TDRSS	Tracking and Data Relay Satellite System
TNT	TriNitroToluene
UHF	Ultra High Frequency
UNCOPUOS	United Nations Committee on the Peaceful Uses of Outer Space
WHO	World Health Organisation
WMO	World Meteorological Organization

Glossary

Accelerator: A machine for increasing the velocity of charged elementary particles and ions.

Anti-ballistic missile: Missile to intercept a ballistic missile during its flight trajectory.

Anti-tactical ballistic missile: Missile to intercept a tactical missile during its flight trajectory.

Apogee: The point in the orbit of an earth satellite that is farthest away from the earth.

ASAT: A weapon to destroy satellites in outer space.

Atmospheric drag: Decelerating force due to air acting upon a satellite.

Boost phase: The first phase of a ballistic missile trajectory during which it is powered by its engines. This phase lasts for about 180 to 300 seconds, and an ICBM would reach an altitude of about 200 km.

Booster: A rocket used to give extra power during liftoff or at another stage of a spacecraft's flight.

Brightness: Source intensity measured as the amount of power that can be delivered per unit solid angle by a directed-energy weapon.

Data: The raw signals including imagery generated from an observation satellite.

Decoy: An object designed to deceive an observer; often a light object designed to look like a satellite or a nuclear warhead.

δv (delta vee): The velocity imparted by a propulsion system to, for example, a warhead relative to a satellite from which it was launched.

Directed-energy weapon: A weapon such as a laser, which destroys its targets by delivering energy to it, at or near the speed of light.

Discrimination: The process by which a surveillance system distinguishes decoys from intended targets.

Early warning: Early detection of an enemy ballistic missile launch, usually by means of surveillance satellites and long-range radar.

Excimer: A pair of atoms bound together only when the molecule is in an excited energy level.

Free electron laser: A type of laser that does not use ordinary matter as a lasant but instead generates radiation by the interaction of an electron beam with a static magnetic or electric field

Fluence: The amount of energy per unit area incident on or absorbed by a target.

Gamma rays: Electromagnetic rays that have wavelengths in the range 10^{-10} to 10^{-13} meter.

Geostationary orbit: An orbit at an altitude of 35,800 km above the earth's equator. A satellite placed in this orbit revolves round the earth once a day, thus appearing to be stationary relative to the earth's surface.

High earth orbit: An orbit about the earth at an altitude greater than about 5,600 km.

Homing vehicle: A vehicle with sensors on board to detect the position or predict the future position of a target. Once the target is detected, the sensors can direct the vehicle to intercept it.

Inclination (orbital): The angle of inclination of the orbital plane of a satellite to the earth's equatorial plane.

Kinetic energy weapon: A weapon that uses kinetic energy, or energy of motion, to destroy an object.

Lasant: A material that can be simulated to produce laser light.

Laser: A device that produces a narrow beam of coherent radiation through a physical emission.

Launcher: A rocket vehicle used to orbit a satellite.

Mesosphere: The ionosphere, one of the four layers of the earth's atmosphere, is divided into two layers called mesosphere and the thermosphere. The ionosphere consists of charged particles and extends to some 640 km above the earth's surface.

Microwave: Electromagnetic radiation with wavelengths in the range 0.01 - 0.3 meters, i.e., in the region between infrared radiation and radio waves.

Neutral particle beam: An energetic beam of neutral atoms, i.e., atoms without net electric charge.

Orbit: The path of a satellite under the influence of the earth's gravitational force; the path is a closed one so that the satellite returns periodically to the same point in the path.

Orbital elements: A set of six parameters defining the orbit of a satellite. These are the right ascension of the ascending node, the argument of perigee, the orbital inclination, the semimajor axis of the orbit, the eccentricity of the orbit, and the time of perigee passage.

Perigee: The point in the orbit of an earth satellite that is closest to the earth.

Phased-array radar: A radar with a beam that can be electronically steered but has its elements stationary. It can switch rapidly from one target to another.

Polar orbit: An orbit having an inclination of 90°.

Radar: A device for detecting targets in the atmosphere or in space by transmitting radio waves (e.g., microwaves) and sensing the waves reflected by objects.

Reentry vehicle: The part of the ballistic missile that carries a nuclear warhead to its target. It is designed to reenter earth's atmosphere in the terminal part of the missile's trajectory.

Resolution: Ground resolution is the ground dimension equivalent to one line at the limit of resolution; photographic resolution is the minimum observable spacing between black-and-white lines in a standard pattern.

Satellite, artificial: A manufactured spacecraft that revolves around a spatial body, such as the earth.

Space mines: Hypothetical devices that could track and follow a target in orbit, with the capability of exploding on command or preprogrammed to destroy the target.

Stratosphere: This layer of the earth's atmosphere contains small but vital amounts of ozone. It extends to about 80 km above the earth's surface.

Telemetry: Transmission of data using microwaves (e.g., housekeeping data, imagery data from satellite sensors, etc.), often using coded signals.

Terminal phase: The final phase of ballistic missile's trajectory, which lasts for about 60 seconds or less.

Tethered satellite: Two satellites, flying one above the other, connected by a rope or a chain.

Thermosphere: One of the two components of the ionosphere.

Tracking: The monitoring of the course of a moving target.

Tracking system: Systems such as radars or optical telescopes, which follow the movements of satellites in earth orbit.

Trajectory: The dynamical path followed by an object under the influence of gravity and/or other forces. A satellite's orbit is a trajectory that is closed so that the satellite returns periodically to the same point.

Troposphere: One of the four broad layers of the earth's atmosphere extending to about 20 km. It contains the bulk of the atmosphere and all its weather.

Warhead: A weapon, usually nuclear, contained in the payload of a missile.

X-ray laser: A laser that generates a beam or beams of X-rays.

X-rays: Electromagnetic radiation having wavelengths shorter than 10 nanometers (10 billionths of a meter).

List of Contributors and Consultants to the Book

Prof. M. Khairy Aly
Helwan Institute for Astronomical and Geophysical Research
Helwan
Cairo
Egypt

Dr. Raoul A. Amat
President
Federacion de Asociaines Nucleares de Americia Latina
C.M. della Poalera 299, piso 15
1001 Buenos Aires
Argentina

Prof. Bruno Bertotti
Dipartimento di Fisica Nucleare e Teorica
Universita degli Studi di Pavia
Pavia
Italy

Dr. Yves Boyer
Deputy director
Centre d'Etude des Relations entre Technologies et Stratégies
CREST
75014 Paris
France

Dr. Ashton B. Carter
Prof. of Public Policy
Acting director
Centre for Science and International Affairs
John F. Kennedy Street
Harvard University
Cambridge, MA 02138
USA

Mr. S. Chandrashekhar
Launch Vehicle Programme Organisation
Headquarters
F-Block, Cauvery Bhavan
K.G. Road
Bangalor 560 009
India

Ambassador Jayantha Dhanapala
Director
United Nations Institute for Disarmament Research
UNIDIR
Palais des Nations
CH-1211 Geneva 10
Switzerland

Dr. Pal Dunay
Associate prof.
International Law Department
Eotvos University
Budapest
Hungary

Dr. Hubertus M. Feigl
Stiftung Wissenschaft und Politik
Forschungsinstitut für Internationale Politik und Sicherheit
8026 Ebenhausen
Germany

Prof. Dr. Dieter Felske
Neustrelitz
Germany

Prof. Donald L. Hafner
232 Rutledge Road
Belmont, MA 02178
Massachusetts
USA

Dr. Bhupendra Jasani
41 Tenterden Drive
Hendon
London NW4 1EA
UK

Dr. Thomas H. Johnson †
Prof. of Applied Physics
Director
Science Research Laboratory
Department of the Army
United States Military Academy
West Point, NY 10996
USA

Dr. S. Rodionov
Space Research Institute
Profsoyuznaya Street 84/32
117810 Moscow
USSR

Ms Isabelle Sourbès
Chargée de recherche au CNRS (IMAGEO)
Chargée d'études au CREST
Centre d'Etude des Relations entre Technologies et Stratégies
CREST
54 rue Boissonade
75014 Paris
France

Dr. Paul Stares
Brookings Institution
1775 Massachusetts Avenue, NW
Washington, DC 20036
USA

Prof. Serge Sur
Deputy director
United Nations Institute for Disarmament Research
UNIDIR
Palais des Nations
CH-1211 Geneva 10
Switzerland

Prof. Ivan A. Vlasic
Institute and Centre of Air and Space Law
McGill University
3644 Peel Street
Montreal, Quebec H3A 1W9
Canada

Prof. C. Voûte
Stokhorstlaan 83
7531 JE Enschede
The Netherlands

Dr. Chen Zhongquing
Senior engineer
Ministry of Astronautics
Beijing
People's Republic of China